Homeowner's Guide to Roofing and Siding

Robert C. Reschke

Ideals Publishing Corp.
Milwaukee, Wisconsin

Table of Contents

Technical Advisor, Frank Pershern
School of Industry and Technology
University of Wisconsin-Stout

ISBN 0-8249-6107-2

Copyright © 1981 by Ideals Publishing Corporation

Published by Ideals Publishing Corporation
11315 Watertown Plank Road
Milwaukee, Wisconsin 53226

Editor, David Schansberg

Cover photo by David Schansberg

SUCCESSFUL
HOME IMPROVEMENT SERIES

Bathroom Planning and Remodeling
Kitchen Planning and Remodeling
Space Saving Shelves and Built-ins
Finishing Off Additional Rooms
Finding and Fixing the Older Home
Money Saving Home Repair Guide
Homeowner's Guide to Tools
Homeowner's Guide to Electrical Wiring
Homeowner's Guide to Plumbing
Homeowner's Guide to Roofing and Siding
Homeowner's Guide to Fireplaces
Home Plans for the '80s
How to Build Your Own Home

Your Working Options

Consider the exterior of your home in its entirety. Look not just at the re-roofing or re-siding work that plainly needs to be done; look at the whole. Do you like the appearance of your home's exterior (windows, entrance, doors, soffit, etc.)? Is your home as energy efficient as you would like? Are you tired of scraping and painting every couple of years? Look at the whole exterior again and plan the renovation carefully so you are sure to get the exterior you want.

Exterior renewal work has many options. You can employ a specialist in one specific area, obtain a well-rounded remodeler for most of the work, or do much or some of the work yourself. Roofing and siding work can be separated into major areas and supplemental jobs that may or may not have to be done, or added jobs you simply wish to carry out. In this initial chapter, the focus will be on your various options.

Preliminary Considerations

Many roofing and siding jobs can be done without building plans and specifications. As long as you're just re-covering the surface and not changing the structure, a building permit and formal plans and specs are not required. Nevertheless, you should still make careful plans on paper for materials, work details, and other information as you think about the job and organize it.

Exterior renewal involves many related details such as chimneys, insulation, adding skylights, removing porches and replacing windows and doors. Make notes and plans about details and certain materials or products; keep track of contractor or supplier contacts; maintain a name-address-phone list for information or service sources. Even simple improvement jobs require some record keeping. The more work you contemplate, the better organized and more detailed these informal plans and records should be.

One of the reasons for records is federal income tax filing. When you sell your home, it will undoubtedly sell, due to inflation and other factors, for more than you paid for it. You must pay income taxes on the difference between the buying and the selling price unless you buy a higher-priced home within a year. Before determining the tax to be paid, you are permitted to deduct the cost of certain improvements, but such improvements must have been performed during the 90-day period ending on the date of contract of sale (closing date), and it must have been paid for within 30 days after closing. You must have records to support your deduction claim.

In addition to your options in choosing contractors you will also have a choice of materials. You'll quickly become aware that there are different grades of quality in different lines of the same type of material. And, you'll discover that many contractors and suppliers don't like to offer you a complete idea of what's available—many prefer to recommend just the one or two grades or products they carry.

When considering the various options, remember your family still has to function during the renewal work, so be prepared to live with the mess and inconvenience.

The legalities For any kind of structural change to your home, you will need a building permit. If a contractor will do all or part of a job, he may offer to obtain the building permit. But the permit is your responsibility and it is you who will be held accountable if work is done on the home without first obtaining the necessary permit.

Visit your community's building department and talk with the building official, the building inspector or some other responsible person in the department to verify your need for a permit. Explain the scope of the work you have in mind, what changes you plan to make if any and find out if a permit is required. If it is, ask about the forms you have to fill out and the drawings or specifications you must submit with the permit application. Permits generally require payment of a fee; fee practices vary from one locality to another.

Your neighborhood may have a development association which must approve exterior changes in all homes in the subdivision, land tract or community. While you are at it, examine your deed. Are there any restrictions covering exterior materials on homes?

Check your insurance. Your home is probably insured with various kinds of coverage, including a liability provision that will cover accidents to members of your own family and also to outsiders. But re-roofing and re-siding may affect the insurance coverage. Ask your insurance company or agent about this and learn just what limitations your policy may have with respect to remodeling accidents and injuries. Check on accidents to workmen on the job. You need to know whether the contractor has proper workmen's compensation insurance so that you will not be found liable for compensating the workmen for injuries.

If your contractor's men or equipment must go beyond your property line, get your neighbor's permission before proceeding with the work.

Financing improvements Most of the time, financing home improvements is entirely up to you. Short-term home improvement financing involves relatively high interest rates and because homeowners are ordinarily good credit risks, banks and savings and loan associations are usually eager to make such loans. Most small roofing and siding contractors will leave financing arrangements in your hands, and some contractors will offer the name or names of lending institutions with whom they have dealt in the past. Always shop for the best financing deal just like you would for any other service or material. There are loan limitations and variations in disbursement practices. Find out the details from the loan officer at the bank before you sign work contracts with any contractor.

What work should you do yourself? In the course of determining what needs to be done, what can be conveniently done at the same time, or what you'd like to include, your efforts should include a task breakdown. Spell out, in a written agreement or contract, the degree or scope of work you want the contractor to handle. Spelling out details in this manner avoids misunderstandings. A second reason for this breakdown is to let you stand back and size up the entire project. You can estimate the time needed for each segment of work. Keeping separate jobs separate does allow you to put them together mentally in the proper slots and time schedule.

Plan and gather information about procedures for all tasks in advance, whether or not you plan to do all the work yourself. In fact, it's an excellent idea to take advantage of contractor's "free estimates" on the work you do think you'll handle yourself. It gives you a more specific idea of how much savings are possible doing your own work. Sometimes, the prices quoted by contractors are surprisingly low. So whenever contractors come in to look at your home and give you an estimate, be sure to ask them for an estimate that covers the main work, plus supplementary work, but make it clear that the proposal should specify the basic cost plus costs for each of the extras.

By separating individual jobs you can determine whether each task is something that fits into your time frame for handling it yourself, or whether it is better handled by a contractor and his crew, who would do the work in a fraction of the time you'd take.

The application of aluminum or steel siding demands a certain handling skill and watchfulness, and when damage is done to their surfaces it is almost impossible to set them straight again or repair them.

Contracting Options

Some states have laws regarding who may work on certain kinds of home inprovements, the laws usually being in the form of business licenses. Such laws tend to control who obtains a license and what must be done to keep the license in good standing.

Types of contractors While there may be exceptions in certain areas to the following generalizations, you'll find over most of the country that some roofing/siding contractors are generalists doing several different types of work, while some are specialists who limit their work to just one or two types of work.

There is no way to determine in advance which type of operator will do a better job. However, a specialist contractor is apt to give lower prices in the field of his specialty, which is sometimes the result of his having a smaller operation with lower overhead. Here are some of the different types of contractors in the fields of residential roofing, siding and modernization work.

- Application Contractor—specializes only in new home construction work doing business with professional home builders.
- Carpenter Contractor—individual carpenter who works primarily on remodeling jobs, exterior or interior.
- Roofing Contractor—generalist doing work on all kinds of roofing and re-roofing jobs.
- Shingling Contractor—specialist who confines his work to application of asphalt or wood shingles.
- Roofing-Siding-Insulation Contractor—does all kinds of home improvements on home exteriors and structures.
- Siding Specialist—usual emphasis is on re-siding work involving metal or plastic siding materials.
- Insulation Specialist—usually applicator of special types of insulative materials, such as blow-in.

Obligations to contractors Never pay a contractor in full until the job is completed. Make partial payments only after appropriate portions of the work have been done, and in making such partials obtain a receipt indicating what portion of the work payment covers.

Many contractors are dependable and adequately financed businessmen. If your contract amounts to a very substantial sum, and one for which you must obtain some form of credit, it is frequently possible to arrange extended payments over a period of several months. Such arrangements should be included in your contractual agreement.

Final payment should not be made until all work has been completed to your satisfaction. Use the final check as a lever to have your contractor live up to the contract and to all verbal assurances of attention to certain details.

Some states have a mechanics-lien law applicable to home improvement work. These laws protect subcontractors and suppliers for materials and labor they furnish to a contractor. If the contractor doesn't pay the subcontractor or supplier for materials/labor, these businessmen can obtain a lien against the property ... and then their payment becomes your responsibility in order to lift the lien. Such an eventuality can be avoided if your contractor will furnish you with a certification that he has paid the suppliers and subs that were involved on your job. This certification can be easily done in most states with such laws by having those suppliers and subcontractors fill out waiver-of-lien forms. This is a common practice in many states for new construction, but the practice has been limited in the home improvement field.

Obtaining contractor bids A bid is an estimate or a proposal. The three words mean just about the same thing. When you call a contractor about some prospective work, you expect him to suggest the scope of the work to be done. He will try to gauge your needs or wants, and then quote a price on the various jobs.

Some contractors or home improvement firms like to give this figure to you verbally; they call it an estimate. Then, when you mention related work and ask if that is included, the answer is inevitably "No" and the estimate goes up.

What you want to obtain is a "proposal." In this written work-and-price-quoting offer, the contractor will outline the work he proposes to do for you, what materials will be used, what assurance he can offer you of a good job; and then he will give you a price. At the bottom, a signature line for you is provided and to accept his offer, all you need do is sign the proposal. Essentially, a "bid" is exactly the same thing. When you go around asking contractors to submit proposals to you on doing some work, you are essentially taking bids.

To obtain the names and telephone numbers of contractors or improvement firms who can give you bids or proposals, there are three principal sources (none of these include calling friends or acquaintances, who are usually poor sources for good contractor information):

● Local building material suppliers or retail lumber yards. These companies often have contractor departments that will give you names of roofing, siding, or remodeling contractors. You're likely to get names of generalists from this source.

● Your local Yellow Pages, if you look in the appropriate categories or classifications: "Home Improvements," "Roofing Contractors," "Siding Contractors." The display ads in these sections often give further guidance.

● Ads in local community newspapers. These often give names of contractors who handle work only within a limited area; look also in the classified listing columns.

Before you call contractors you have some further research to do first.

Your renewal program needs to be developed further before your contractor contacts. Not every contractor will make the same recommendation. You need to have a pretty fair idea of your home's condition and status so you can better discuss details with them. And also to better judge how well they seem to know their business. You want to have sufficient knowledge about the materials to be used and the supplemental jobs that may be necessary or desirable, so that you can intelligently discuss these possibilitites with contractors. And your orientation to the jobs that might be contracted must be comprehensive, so that you avoid signing an initial contract and then decide later that you want to make changes. Such indecisiveness can be costly.

Before making any decisions on the various contracting options that may turn up as you call in contractors to talk over the needed work, prepare yourself. Check your home's condition, and gather information that will permit you to formulate suitable roofing, siding and exterior renewal plans.

Making a Structural Checkup

You know how your dentist examines your teeth when you go in for a checkup visit. He carefully goes around, tooth by tooth, with his little mirror and probing tool. He views different sides of each tooth. With the probe he tests known spots where decay gets a start. He goes down around the low sides, sometimes below gum line ... probing, probing ... and sometimes substituting a small knife to scrape away plaque at some spot so that he can see below. Then, he seeks out hidden trouble via X-rays.

Unfortunately, no X-ray machine has yet been developed to spot below-the-surface troubles on a house. But otherwise, when you're considering the possibilities and needs for re-roofing or re-siding your home, your approach to determining those needs should parallel the dentist's technique; visual checks of all sides; probing and testing of the condition of wood, asphalt and metal; sampling attempts to see what's hidden and below the surface; a thorough search of common trouble-prone areas.

The checkup suggested here refers to something more than a few glances at structural framework. Sheathing is also a part of the structure, and its condition should be noted and in places tested. While roofing and siding work deal primarily with surface coverings, the under-surface base upon which they rest and to which they are fastened is also part of the "structure."

Your roofing-siding or exterior checkup may be initiated for any number of reasons relating to your present home, or it may be for a house you are considering as a purchase.

It really does not take a building expert to recognize conditions that are inadequate. As you proceed in the manner described, you will discover things that earlier, casual glances never revealed, and this is partly because you weren't sure what you were looking for.

However, if after making your inspection and checkup rounds, you still feel unsure as to whether or not certain work needs doing, then you may need the assistance of an outside expert. There are, in nearly all metro or urban-suburban areas, private building inspection men or organizations. These are individuals or chain representatives and appraisers who, for a fee, will give you an accurate picture of a home's condition. Most have engineering or architectural qualifications and are much more reliable than that brother-in-law or cousin who "knows something about building."

In making your inspection, try to cover all areas under consideration for renewal work. Lower-sloped roofs can be walked on for inspection, but keep the walking to a minimum since it can in itself be damaging. With steeper roofs, avoid unnecessary climbing around; sometimes inspection of roof surfaces and chimney flashings can be done from dormer windows using binoculars for close-ups.

In using a ladder, observe some safety rules. Place the ladder on a firm base and at an angle of 65-to-70 degrees. An extension ladder will be required for two-story homes but a good quality stepladder may suffice for inspection of a one-story home.

Don't attempt to reach out very far sideways when on a ladder. Always climb and descend facing the rungs or steps. Watch out for ladder feet settling down in soft earth under your weight. Wear soft-soled shoes on ladders and when moving about on roofs.

When you see slight indications at the surface of possible trouble, use a screwdriver to probe below the surface. Do it carefully because if there are soft spots or decayed wood below, it digs out easily and rapidly. Test various critical areas for firmness. Mostly what you're looking for on the home's exterior is evidence of moisture damage.

Moisture—The Enemy

The durability of wood fibrous building materials is nearly always lengthened and strengthened by absence of moisture. Sometimes moisture can be controlled. One way of controlling the presence and buildup of moisture is by ventilation; wood materials that do occasionally become moist seldom decay if they quickly dry out again through good ventilation.

When wood materials do become alternately moist and dry, their dimensions change—larger because of swelling with moisture, smaller because of shrinkage when dry. This expansion-contraction in some cases may allow moisture to seep into the cracks between materials. Good caulking in such joints or openings will remain flexible, and bend with the expansion-contraction movements.

Moisture accumulations in attic spaces and within exterior walls can reduce the thermal resistance value of insulating materials, and thus increase fuel consumption and cost. Entrapped moisture on many painted exteriors is the principal cause of blistering and paint peeling, just as it is also the main villain in blistering of asphalt roof coverings.

Further evidence of moisture problems can be seen in the buckling and ridging of built-up tar-and-gravel roofs where underlying asphalt felts have become damp. When insulations become continously wet, there may be a cellular breakdown in the rigid board types and organic insulations will tend to decay. Mineral or fiberglass types may exhibit binder failure. Moisture presence at one time or another may produce surface discolorations. Be skeptical of irregular brownish lines on lumber and plywood which are quite possibly caused by the spread of moisture.

Often when wood surfaces appear to be dry, they actually are not. You can test a wood surface by using a coin, key, or screwdriver to scratch across the wood grain. If the scratching produces slivers of wood that break loose, the wood is probably fairly dry. If, however, the scratching doesn't break loose any fibers and simply results in an indented mark, the wood is probably quite moist.

Moisture evidence on exterior walls often takes the form of darker areas where surface dirt films tend to be greater and more difficult for rain wash-off. Moisture must also certainly be suspected underneath areas or cracks that develop natural growths such as moss or fungus. White discoloration areas on masonry might raise suspicions of the tightness of the masonry joints, and the probable need for tuckpointing or a waterproofing surface treatment.

A leaky roof makes it obvious that the source of your moisture is in the rain and snow outside, and the way to correct the difficulty is to provide a better roof covering to stop the entry of moisture.

Some moisture problems arise from flow in the opposite direction. A great deal of moisture is generated by normal cooking, dishwashing, laundry, bathing and cleaning activities within the home. This becomes moisture vapor in the interior air and the moisture-laden air tries to find its way outward through the ceiling and walls of a home. Without a vapor barrier on the inside, over the wall insulation, moisture vapor reaching outer wall or ceiling areas will condense in cooler weather and accumulate on insulation, wood members and surfaces.

While it would mean complete removal of interior plaster or wallboard to apply a vapor barrier to a home lacking one, there is another way to control such condensation and that is by means of adequate ventilation for attic and eave areas. Even in homes having vapor barriers, moisture vapor can still penetrate to some extent into these areas through scuttle doors, attic stairways, cracks or openings in the barrier or other ways. Good ventilation is the control measure.

Evidence of the shrinkage of wood fibrous materials is often seen in the form of cracks or openings.

The good news about shrinkage cracks is that they are excellent indicators of thoroughly dry conditions. The bad news is that shrinkage cracks in the wrong places often provide channels for moisture seepage and entry from outside. The control measure for most cracks is suitable filling or caulking. What you need is the kind of caulking material, and there are several types, that remains flexible despite age even though forming a surface skin that becomes hardened and smooth.

Checking Roof Conditions

If you have a roof leak that makes itself known by ceiling drips, you may sometimes be able to pinpoint the approximate point of leakage on the underside of the roof sheathing, then go up on the roof at that location and detect what may be causing the leak. But chances of doing that are slim, because most roof leaks meander. They start at a certain point and the water coming in finds its own irregular and wandering path across and downward. Where it eventually comes through the under-surface of the roof boards may be quite some distance from its point of origin, and where it comes down on the ceiling may be yet another good measure from the place where the water channel can be seen on the roof boards.

There are times when a nearby chimney, plumbing vent or roof ventilator with disturbed flashing may be the source of leaks. One big trouble with leaks is that you often need to look for them while it's still raining, or shortly thereafter, in order to pinpoint the source.

A common area for roof leaks around flashings is near a roof penetration such as a chimney, plumbing vent pipe and roof ventilator. Still, the visual spotting of faulty roof flashings is difficult, too. Often a homeowner having roof leaks will not be able to locate the source but will simply have to judge by the general condition of flashings: any looseness, any nearby cracks in material, any separated openings in the flashing laps or folds, and the roof needs work.

Although not always moisture-related, check asphalt shingle roofs for tabs that appear to have been loosened (often by wind lift); look for any loose shingle areas for evidence of loosened roofing nails.

A single leak in a roof may be due to an unusual circumstance such as a heavy tree limb falling on the roof. A persistent leak over a certain area may indicate there is more than one point of moisture entry; this, or leaks that move from one location to another, indicates a serious need for re-roofing.

Not all moisture penetration difficulties show themselves by the accumulation of enough water to flow down and become a leak. Moisture may ac-

Sharply modern is this change from the typical turn-of-the-century old home. New front windows, old porch removal, and a contemporary addition helped bring about a startling new look. Dramatic skylights and new

entry windows provide a pleasant interior, while new roofing and vertical vinyl siding plus new vinyl soffits give the entire home a fresh appearance. Drawings courtesy of Bird and Son

cumulate in the building materials and on their surfaces causing long-term damage without ever reaching the ceiling level. Your inspection should, therefore, include a close look at the underside of your roof, and your probing should focus on areas that show any discoloration or "water marks" that suggest moisture presence at one time even though the material is dry now.

On roof surfaces around chimneys, vents, and other places where roof flashings occur, you'll want to check for any apparent displacement of flashing material and for water-trapping crevices. The sure test of a moisture problem in such areas is a softness or sogginess of the roof boards or plywood sheathing just below the roofing material. Test with your screwdriver at strategic points where the flashing or roof material can be folded back or lifted slightly to test, and then pressed back into position.

It is along the roof edges that you are most apt to discover evidence of underlying structural damage resulting from external rather than internal moisture. Eaves are especially subject to decay because of a nearly constant presence of moisture. Sometimes, the softness of wood members that support fascia and soffit materials can be almost sickening, as you feel the screwdriver encounter almost no resistance where there should be solid and firm wood.

While the eaves and rake areas may be more prone to moisture damage, they are also easier to repair and get back into suitable shape than other areas of the roof. Damaged rafter ends can be supplemented by short new lengths nailed to the old rafter sides. Damaged lookout members (those pieces running from rafter end to walls to support the soffit material) are quite easy to replace; if there has been long-term moisture in these areas, it will probably be evident in the poor shape of the soffits and fascia,

which can also be replaced with more durable materials.

Poor condition of your roof guttering is usually due to rusting. But often a contributing factor is the accumulation of grime and debris in the gutter and the improper or uneven sloping of the gutter due to faulty suspension. Such irregularity in gutter slope may be caused by decaying roof edge or fascia boards, or even by rotting rafter ends to which the guttering may be spiked.

On old asphalt roofing look for evidence of roofing materials drying out. One excellent indication is in the looseness of mineral granules. When asphalt-impregnated felts dry, the asphalt tends to become brittle and lose its grasp on the embedded granules. If granules have been lost over many areas of the shingles, it's a good bet your roofing has worn out and needs replacement.

Other indicators on roof surfaces are hairline cracks in the materials, upward curls along shingle edges, or cupping of shingle tabs. Obviously when shingle tabs are missing or are creased, giving evidence of wind lift, again it is time to plan re-roofing.

On wood shingle roofs you are not likely to find blisters or dryness. Instead, look for moisture accumulation spots in shingle crevices. Cupping of shingles indicates probable alternate soaking and drying out. Test suspicious spots with your screwdriver—almost any spot or area that is quite dark in color is a suspect area. Discoloration stains on wood shingles may be evidence of places that trap moisture during rains so that afterwards water runs off slowly and creates the discoloration.

Siding Conditions

Less probing but closer visual inspection is likely to be required in your checkup on exterior walls and

Adding a cozy study on the ground level gives this old two story a much newer appearance. The new roof line is extended to provide a cover for the front entry. Heavyweight roof shingles add a rustic touch. Re-siding in the

8-inch width adds better horizontal lines, and provision of new shutters and roof gutters plus a large new picture window are worthwhile improvements.

other exterior areas. Paint problems are usually self-evident in the form of peeling and scaling. This, too, may have a moisture cause but scaling can be the result of normal deterioration, or poor quality paint that has been used in the past combined with moisture.

Whatever, the homeowner is apt to be much more interested in learning about the solution rather than the problem and its possible causes. For most homeowners finding themselves with exterior walls in bad shape, the most suitable and best-all-around solution is a re-siding job using materials that come prefinished and need minimal care on an annual or biannual basis.

One structural difficulty with exterior walls that an owner should try to check is the building's alignment, horizontally and vertically. This can be done visually by placing one's eye in the proper position. For example, you can easily squint to detect any uneven settling that has occurred along a building wall, by squatting down at a building corner until your eye is looking along the bottom edge of the present siding material. If this edge is straight, your building has probably settled normally; if there is curvature in the bottom edge, your trouble is more likely to be due to uneven foundation settling than to improper siding application.

The reason a thorough check for alignment should be made on all building sides and corners is that it will be relatively more difficult to obtain a good re-siding job if the existing siding and trim lines are askew or much out of line. It will probably mean extra expense of some sort to accommodate or adjust to such irregularities.

On older homes, windows, doors, and trim can be troublesome. Paint layers can become extra thick and uncountable. Surface irregularities are wide-spread. Cracks, fine hairline cracks and crevices prevail. These are good indications that your door-window frames and trim should either be recovered or replaced in conjunction with re-siding work. A re-sided home with the old window frames partly concealed may look worse than it did before if nothing is done to improve the window, door and trim condition. For many older homes one excellent though not inexpensive answer lies in replacement windows and entrance doors that accomplish three jobs at once: eliminate the maintenance problem almost completely, contribute to a much better home appearance, and provide energy-saving value because of their insulative properties.

Soffits, gable-ends, eave and gable ventilation, roof overhang appearance, surface condition of exposed portions of foundation walls—these are some of the other external areas of a home which may need reconditioning in one form or another.

By no means least, though dealt with last, are the various external appurtenances that a home may have. Front porches, side porches, rear stoops, guest and family entrances. Very often these home attachments when first built were given insufficient attention. A common porch fault is the settlement of porch pier foundations irregularly or to a different degree from the settlement of the house itself. Under-porch areas often accumulate moisture or show evidence of rotting long before foundation portions of the main home structure.

For many of today's lifestyles, the old-time roofed-over porch is an anachronism. For many homeowners whose use of a porch is limited or nonexistent, the logical answer when re-siding or remodeling a home's exterior is to wreck it—remove and replace it with a new facility that will better serve your purposes and at the same time bring about a more contemporary look. For side porches or rear porches, consider replacement with an elevated-to-floor-level wood patio deck. For home entrances, consider a more durable concrete platform and steps with ornamental iron railings that help set off a new entrance door.

Most porches begin failing from the bottom up. Foundation piers are difficult to repair or replace while the porch remains in place. From an appearance standpoint, nothing on the home's exterior quite dates the home as much as its porch, or porches.

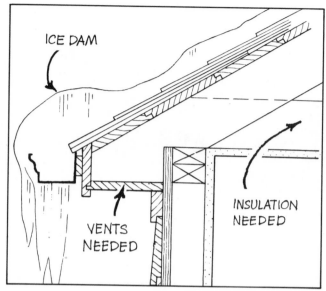

Attention to details in exterior renewal are seen in this renovated northern Michigan home whose eaves and roof valleys have been protected against snow-ice accumulations through the use of sheet metal. The metal-surface eaves allow snow-ice to slide off thus preventing a build-up called an "ice dam" which can back up under shingles and cause roof leaks. In less severe snow areas, ice damming can be prevented with proper attic space ventilation, and the provision of adequate ceiling insulation. Low-slope roofs need the added protection of cemented-down roofing felt under the lower courses of asphalt shingles.

Inspection of old shingles may reveal breaks and loss of granules from the surface, as indicated above. Replacement is needed when such damage extends over much of the roof's area. At right, shingles are cupped or curled and should be removed entirely before re-roofing.

Finalizing Your Plans

Your planning work should cover four types of details:

- Drawings to scale if work involves alterations, otherwise sketches to indicate how you think work should be laid out, re-designed or handled.
- Specifications giving work installation details if the work involves a building permit application; otherwise a file of manufacturers' product literature for type of work being done or materials to be used.
- Notes that relate to sources of supply of materials and contractor services, notes on prices and delivery factors.
- Schedule outline to be developed, with a timetable indicating order of work and promised dates of starting or completion.

This collection of planning details should be started as soon as your first contacts are made and should continue as you gather more information and reach the decision-making stage as to what work will be done and which methods will be used. Each of the four areas of planning depend somewhat on other factors. A manufacturer's brochure will give you general information which you might wish to supplement with a sketch of your own, in order to explain better to some contractor or supplier how you wish to employ the product in your home.

Manufacturers' literature pieces on materials or building products usually give all the details on dimensions, and such things as color choices, but they seldom give prices or delivery information. These you'll have to inquire about from the supplier and take notes for your planning files.

If your exterior renewal work is going to involve more than just one or two simple jobs, a written timetable or schedule is important. You can set it up on one or more large sheets of paper with calendar time along one axis and job description along the other axis. This will permit you to keep track of when contractors are due to come in to work, and their likely completion date, so that work preceding, subsequent to, or dependent upon their completion, can be better visualized.

Such a calendar-type schedule is particularly useful when you are planning to fit the jobs that you will handle yourself, while a contractor (or perhaps more than one) handles other specific jobs.

Scope of the Work

As your plans begin to jell on different exterior renewal jobs, you should set down on paper in brief form what each specific job entails. After some investigation and talks with suppliers or contractors, these ideas on just what work needs to be done may change, but by the time you're ready to either sign a contract or begin a specific job yourself, your notes should tell you item by item the work to be included.

Don't think that your notes and files and the schedule will provide all the answers. They won't. But they will definitely help in making decisions.

Inevitably with any remodeling or modernization job, it turns out there are more little nitty-gritty details to deal with than anticipated. Your files, your notes, and your sketches are the way to put a handle on these little details. They are also a device where tentative or optional items can be fitted in for consideration and easily put aside later if you rule against them.

Opportunity to Modernize

Modernize simply means "bring up-to-date" and does not mean giving a home a modernistic, futuristic, or other far-out appearance. When you plan re-roofing or re-siding work, it's a very logical time to think in terms of modernizing. And the reason for this is that too often, a re-roofing or re-siding job by itself does not significantly improve a home's value. True, better surfaces, more durability and less maintenance ensue and are sales factors later, but the home could still have a dated or old appearance. So, consider your house in terms not just of resurfacings but rather what changes can be made in the exterior without undue expense that will really better the home's appearance, and thus materially upgrade its value.

Go slowly on this. The old worn condition may look bad and the bulky, uneven porches may be an eyesore—but proceed with care. Many older homes have architectural lines that are well worth preserving. In fact, if your house has an exterior that is identifiable as being authentically of a certain period's style of architecture, it may be more appropriately brought up-to-date by a blend of restoration and modernization. There are contractors in many areas that specialize in pure restoration work on older homes, but this type of treatment is apt to run into much higher costs than for simple remodeling or renewal.

A very desirable goal on many not-too-old homes is just to achieve a greater simplicity of appearance. Many homes can be improved by reducing complexities, keeping different exterior materials limited to no more than three types of materials.

If you wish to upgrade your home but believe you need professional assistance to attain good results, scout the locality for architects and home designers that are experienced in residential remodeling work. The drawing or sketching work by such a designer, in the form of new elevations, is not apt to cost any more than roughly 10 percent of the exterior improvement total cost and this could very quickly be earned back by the increased market value of the redesigned home.

Often a home's best appearance results when it looks like a unified cluster of its parts or masses, rather than an assemblage of varied parts. Windows, preferably, should be of the same type and style. There should be a composed look, and entrances deserve something special to lend them an inviting image. Watch out for emphasizing appendages or supplementary elements.

Re-siding problems sometimes crop up because a home's original horizontal or vertical lines are changed or modified in some unnatural way by the new siding. Siding materials usually come in either horizontal or vertical types. If you cover old horizontal siding completely with new vertical siding, your home will have a completely changed look, probably not at all in keeping with its basic lines. By the same token, however, many homes had original siding applied without regard to architectural lines and the change may be a step back towards proper proportions.

While considering a re-roofing job it is a convenient time to also think about updating the roof's appearance by removing dormer windows, or adding partial hips, or changing a certain roof slope. In keeping with a residential design trend now occurring, there are a growing number of home conversions to mansard and gambrel roof styles, both of which incorporate a shallow upper roof slope with a steep lower slope. Plain, no-overhang two-story homes that have low-slope roofs are relatively easy to modernize in this style by adding simple framing to the exterior siding on the home's second floor.

Other reasons for changing a home's exterior may stem from what you wish to take place on the interior. Obvious examples are the adding of a dormer window to provide light and ventilation for an attic-room-to-be, and replacing one or two old windows on the first floor with a sliding glass door to give access to a new patio deck. Or carefully weigh the possibility of updating your home's attic ventilation, or the usefulness of skylights in certain locations, or adding a freestanding fireplace.

But bear in mind that work involving structural changes of any kind will necessitate obtaining a building permit and observing building code regulations for modifications or additions.

Many home remodelers will provide the necessary plan-drawing and specification-writing service needed to obtain permits. What generally is needed is: (1) reasonably accurate drawings to scale, blueprints or white prints which show structural before-and-after changes; (2) specifications which indicate materials to be used, their grade or quality, and any load strength or other property which may be pertinent to the changes being planned. If you are planning to handle much of the alteration or remodeling work yourself and are unable to work out the drawings and specs, you can probably find a designer-draftsman or architect in your area who, for a reasonable fee, will put down on paper suitable drawings and specs for permit submissions.

Gathering Information About Sources

Home modernizing ideas that are presented in home-garden or shelter magazines often contain ads or listings of manufacturer sources of roofing-siding and other exterior improvement information. Manufacturers are fairly conscientious in responding to these consumer inquiries and will frequently send back to the inquirer not only appropriate brochures or literature but also the names and addresses of dealers, installers, or retailers who handle or stock their products.

This is just one way to gather modernization information. Other ways include personal visits to building supply or home centers and phone calls to companies listed in local community directories, newspapers, or Yellow Pages. But a homeowner should understand that roofing, siding, and home improvement businessmen are not literature-oriented and they are not prone to sending information in the mail. They would much rather you make personal contact by visiting their office or display room and by letting them visit you in your home. Most home improvement jobs are contracted for and completed without the homeowner ever seeing any printed literature about the materials involved. But this picture is changing as new federal rules are being observed with respect to warranties on home improvement products.

Your best bet to obtain reliable information both in the form of sales literature and other details is to make personal visits to both suppliers of materials and to the business places of contractors or improvement specialists, where samples may be inspected and you can have a preliminary question-answer session with the people involved in supplying or installing the products.

You may be surprised to learn that some building product manufacturers do not wish their outlets to sell materials to people who do their own installations. They will send their authorized dealer or application specialists out to see you but these businessmen are likely to quote only on an installed basis. This applies to a few roofing materials, but the practice is more widespread among manufacturers of aluminum and vinyl plastic sidings; but you can find a way around such manufacturer policies. It is quite possible, in fact likely, that you can purchase the materials from the same wholesale supply sources from which the authorized dealers and application specialists buy their materials. This probably means a cash purchase and that you must place the order in person. Few suppliers or wholesalers will turn you down when you walk into their shop or warehouse, ask knowledgeably about materials they have in stock and offer to pay cash.

In making your tours of supply or service sources, be aware that the selling of home improvements is not all that different from the selling of TV sets or automobiles. You're going to bump into exaggerated or angled claims by product manufacturers in their literature and you're going to encounter salesmen exceeding and misrepresenting the facts about both materials and their installation.

Very often, with a specific product, the benefits and advantages are couched in what is essentially anti-competitor language. Witness the following list of product claims made by a manufacturer of hardboard siding. The parenthetical comments are the author's:

- Impervious to impacts that will dent (like metal siding).
- No temperature expansion or contraction noises (like metal).
- Won't split or crack (like solid wood).
- Doesn't attract or conduct electricity (like metal).
- Concealed nailing (not possible with solid wood).
- Won't pop out to expose sharp edges (like metal).
- Won't interfere with TV reception (like some metals).
- No brittleness, will not break (like plastics in cold weather).
- Adds insulation value, conserves energy (but not by much).
- No corrosion (like some metals).

It's easy to see that the advertising copywriter was instructed to focus on all the competition's apparent shortcomings, real or fancied.

In contrast to the above negative list of claimed advantages by a building product manufacturer, you will occasionally encounter folders or factsheets which do offer a reliable guide to a product's real worth. Here is a paragraph quoted from a technical bulletin issued by one producer of aluminum siding, a bulletin that was available to prospective siding customers of one of the company's installation specialists (and presumably from most others also):

A typical value used with aluminum siding is that given in the *Handbook of Fundamentals* issued by the American Society of Heating, Refrigeration & Air Conditioning Engineers that the average R value of aluminum siding with ⅜ inch thick foam insulation covered with aluminum foil is 2.96. This compares with R = 11 for full 3½ inch thick fiberglass insulation for stud walls.

Now, you may not know the exact meaning of these values offhand but it's something you can question and verify. And the paragraph presents an authoritative source for the information. This is the kind of dependability a homeowner should seek out in manufacturer's literature; although unfortunately, too many manufacturers seem to be convinced that exaggerated claims equal better salesmanship.

Home Improvement Loans

Many home improvements are probably paid for in cash, or the homeowner takes money from his savings to cover improvement costs. This is certainly the case when only one basic type of improvement is involved.

More extensive renewal work, remodeling and new additions can, of course, run costs up quite a bit. At the present time, the financing lid or maximum for home improvement loans will be $10-15,000 depending upon the type loan, the lender's policies, and the borrower's qualifications.

There are a variety of different kinds of loans for home improvement purposes, but there are variations from state to state and one locality to another. In periods of tight money, loans may be difficult to arrange and during such periods lenders tend to favor the less risky borrower. Some loans are made without collateral or security, while others do entail such provisions . . . usually as a lien on the property.

Lenders compete with each other like most other businesses. While there are state laws and federal regulations that control certain factors, this does not mean that all lenders offer approximately the same deal. By no means. So, if you reach the point where you want to borrow money for home improvements, plan to shop around for loans just the way that you would for contractors or for materials.

Don't overlook the variety of sources that you may have available to you. In addition to commercial banks and savings banks, there are also savings and loan associations. And credit unions. You can borrow from insurance companies against your life policy, or contact commercial finance organizations. Contractors and improvement firms sometimes have working arrangements of one kind or another with a lending institution that can be a loan source.

Relatively few homeowners would want to refinance their home through a new mortgage that would supply the improvement funds, because this would probably mean going from a lower-interest mortgage to a higher one. But there are some mortgages that are open-ended and can be added to, though probably at a higher rate than the old interest rate. So, with several possible loan sources, your approach should be to several types of lenders to ask specifically what each offers.

A few years ago, Congress enacted a law which set up guidelines for lenders and credit extenders which made it mandatory for them to provide prospective consumer borrowers with adequate and appropriate details about the loans they were offering, or the installment credit that was being given. One of the objectives was to allow consumers to make suitable comparisons between competitive lending offers. Loans for home improvements, of course, are included under this law.

What the lender must give you is a certain cost figure. He's free to quote you in various ways but must, at the same time, advise what those cost charges amount to in terms of an annual percentage figure. This figure is called the APR, short for Annual Percentage Rate.

Another factor related to truth-in-lending involves home improvement jobs that have package financing. The contractor either has a lender with whom he is affiliated or to whom he refers the customer, and with the loan arranged, the contract is signed. By law the customer has three days in which he can decide whether or not he really wants to continue the contract he signed. He can cancel it within this period by notifying the contractor in writing. Remember that this applies only if your home improvement work is to be financed by an instrument in which the home acts as security for the loan.

Due to differences in state law there are wide variations in the handling of loans in different areas of the country. Loans may be obtained from federal savings and loan associations, credit unions, commercial banks, urban banks, and other lending institutions. Interest rates, maximum amounts, terms, collateral, and methods of repayment change often, therefore, the institution should be called to get the current rates and conditions.

In some localities, second mortgages are controversial. There are some financial men that are outspoken in their criticism of the practice of making second mortgage loans. They say second mortgages are risky for the homeowner because of the possibility that it might violate the first mortgage contract. You may wish to consult a lawyer in your area for his opinion on second mortgage loans before you sign a contract involving this secondary lien on your property.

You may wish to check into tax provisions in your state with respect to home improvements. On the federal level, of course, the interest that you pay for home improvements is tax deductible on your income tax return.

Assuring Homeowner Satisfaction

The home improvement field, like certain others involving consumer products and services, has its share of rip-off artists. Not many, but enough to make customer complaints fairly frequent.

The high incidence of home improvement complaints has led to the formation of a set of National Standards of Practice for the Home Improvement Industry. These have been worked out through joint efforts of the BBB Council, the National Home Improvement Council and the National Remodelers Association. It's interesting to note that the above standards are divided into three parts: advertising, selling the product or service, and taking care of the customer after the sale.

From the work that BBB offices around the country have done in investigating consumer complaints against businessmen, and in serving as mediators or arbitrators in court cases, there are two notable sidelights: (1) in many complaint cases the consumer arguments are not really factual, and the businessmen involved have acted ethically and legally; (2) the most common characteristic in a majority of consumer complaints is a lack of communication that leads to a misunderstanding.

It is clear that most complaints arise from the procedures that home improvement businessmen employ in obtaining the customer's order, rather than from poor workmanship or use of inferior materials. According to citations by the Federal Trade Commission resulting from consumer complaints of firms selling home improvement services, these are some of the types of misrepresentation:

- a sales gimmick that encourages you to sign a sales contract now;
- a time limit for obtaining a special discount;
- price reduction offered for vague or lightweight, nonvalid reasons;
- sales talk that is heavily anticompetitor in nature;
- loose use of verbal-guarantee type language;
- offers to finance cost through credit which involves the use of a nonrecourse instrument.

Two of the most common sales gimmicks used to obtain home improvement contracts play upon the homeowner's vanity and gullibility. The offer of a discount because the contractor can use your home as a model or sample to show other prospects is a gimmick that hopefully has run its course, because many homeowners are now wary of such tactics. But the offer to give you a special deal because the contractor's men are working on a couple other jobs in the neighborhood still arouses homeowner interest. Beware of the "convenience approach" in which the contractor simply drops in on you; under no circumstances give any home improvement contractor or specialist any money in advance as a deposit.

Selecting Reputable Contractors

The National Home Improvement Council is comprised of building product manufacturers, contractors, dealers, and remodelers in more than 30 regional chapters located primarily in larger cities around the country.

The chapter organizations are small in proportion to the number of home improvement businessmen operating in the chapter area, but these groups do tend to work on the consumer's or homeowner's behalf by making the remodeler or application specialist a more responsible businessman. The NHIC and its chapter officers encourage member adherance to sound advertising guidelines, and also promote other useful activities such as bonding programs and setting up of channels for settlement of consumer complaints or disputes.

Among other things, the NHIC has developed a code of ethics for improvement contractors. From it, homeowners can better discern between ethical home improvers and the unreliable type.

(1) Encourage only home improvement projects that are structurally and economically sound;

(2) Make all advertising statements accurate, and free of the capacity to mislead or deceive the consumer;

(3) Require all salesmen to be accurate in their description of products and services;

(4) Write all contracts so that they are unambiguous and fair to all parties concerned;

(5) Perform all work in a manner compatible with recognized standards of public health, safety and applicable laws;

(6) Fulfill all contractual obligations.

These are fine rules. They're probably observed in a greater degree than one might suspect. But there is no testing of their observance nor is there any other enforcement procedure. So the home improvement purchaser must still make up his own mind about the improver's reputability and probable performance.

Manufacturer connection The factory-to-you approach has little meaning in the re-roofing, re-siding and exterior renewal field. The man who claims it is pressing for a job and trading on somebody else's good name.

Place of business A shiny showroom or display area in a reasonably nice building on a main thoroughfare is more an indication of ambition to get more business than it is of working competence or reliability. Untold numbers of first-rate application contractors and specialists have their offices in their home and their wives as phone-answerer and bookkeeper.

Financial references Consumer protection agencies suggest you ask for financial references. Don't waste your time. These are usually privately operated firms and you won't be able to find real financial information even if you have the names of their banks or the lending institutions with whom they work.

Business bureaus In many cities, the Better Business Bureau can give you information about whether there have been complaints about the improvement firms you're considering, and whether such complaints have been settled. Calling the local Chamber of Commerce is of little help and in most cases a waste of time.

Proposal forms The commonly used form of contract by a majority of home improvement contractors is usually satisfactory where the improvement work is limited to one type of operation. On extensive remodeling work, a proposal form usually has too little space to show in detail what work is or isn't included, and you would be better advised to have a lawyer draw up a contract or to include as part of the contract prepared remodeling drawings and specifications.

Material trade-names Where possible avoid proposals or contract agreements that describe the materials to be used as "name-brand" or "best quality." Ask that specific trade-names of materials be given and that you be advised in advance of what warranties are provided on the materials.

Shop around Ask three home improvers to submit proposals: talk to each to see how willing he is to provide what you have in mind; how patient he is in explaining the procedures; what suggestions he offers. Ask about timing and when the job can be completed. You are not necessarily seeking the lowest bid or price but determining which contractor or specialist will do justice to the job.

Previous customers The very best evidence of what a contractor or specialist can do for you is what he has already and recently done for someone else. Ask for names and locations of two or three previous customers to inquire or make arrangements to visit the previous job and to talk with the owner.

Warranties—A New Ball Game

In late 1974, Congress enacted the Magnuson-Moss Warranty Act providing for the establishment of rules and regulations to be issued by the Federal Trade Commission covering the use of warranties or guaranties on a broad range of consumer products. The initial three rules under this Act became effective in January 1977.

The aim of this legislation, and its subsequent FTC rules or regulations, was to clear up consumer confusion with respect to product warranties and to make product manufacturers more uniformly responsible for defective products and for resolving consumer complaints about the products being purchased.

What Congress had in mind was a two-fold goal: (1) organization of the rules of the warranty game so as to stimulate manufacturers, for competitive reasons, to produce more reliable products; and (2) to give the consumer enough information and understanding about warranties so he could look to the warranty's duration and service aspects for an indication of product reliability and performance.

Virtually all home improvement products, including roofing and siding materials, come within the bounds of the Warranty Act. The three FTC trade rules thus far issued cover details concerning three aspects of warranties.

(1) The warranty-issuing manufacturer must disclose the terms and conditions in a written warranty in simple and readily understood language; (2) Sellers of consumer products must make the manufacturers' written warranties available for easy inspection by prospective purchasers prior to the time they commit themselves to the purchase; (3) When a product manufacturer chooses to establish a method of handling complaints and settling disputes, he must follow a set of FTC requirements including disclosure of details to purchasers about the availability of this program and how it works.

Learning warranty provisions and after-sale service on the products and materials could prove to be one of the most significant steps you'll take in the entire exterior renewal process.

Your Roofing Needs

What puts a homeowner on the way to a re-roofing job? Usually it's a leak that recurs in the same place, rather than a decision to spend money on new roofing as a part of an overall exterior upgrading.

Roofs do wear out. If asphaltic, the material becomes thoroughly dry and sheds mineral granules. It may begin to look bad long before leaks develop. Often, the appearance of the old roofing looks its worst when the home has just had a re-siding job or other exterior remodeling. However, homeowners often become used to seeing the worn-out roofing and only the neighbors realize something should be done with it.

Procrastination and delay in re-roofing a home is understandable. Its cure begins largely in a close inspection of the home's roofing materials and also in the status of other roof edge materials ... the roof gutters, fascia and soffits. These are places where damage to the construction is most frequent—damage that stems from the roof's ability to handle water drainage from rains, and meltings of snow and ice that collect on the roof.

If you suspect your home needs re-roofing, the convincing arguments either way are relatively easy to find right there on the roof and along its edges.

A thorough inspection of roof conditions should include close-up visual checks on all sides or slopes. At this point you may ask, "Why should I go to all this trouble of careful inspection and try to learn about the roof's condition? Why not just call in a couple of roofing contractors? Let them inspect and tell me what needs to be done."

Good question. Good, that is, if you believe you will get a fair shake and a fair price from any two or three roofing contractors listed in the Yellow Pages. However, for the dollar-conscience homeowner who has improvements in mind other than just a new roof and who may be trying to decide whether to have a contractor do it or to tackle the job himself, it will pay to learn some of the basics about re-roofing. And a first-hand close-up inspection is a good way to start.

Roofing Terminology

In order to know what you're talking about when dealing with re-roofing contractors or material suppliers, you need to understand a little of the typical roofing jargon. While no doubt there are local variations, they're usually found only with certain specialized materials. Here, the discussion focuses on just a few general terms.

The most common word encountered in roofing circles is "square," a noun. A "square" is simply 100 square feet: an area of roof measurement. When you order or buy roofing materials, you obtain sufficient to be applied on so many squares of roof. Roofing materials are often packaged or bundled so that a certain number of packages will cover a square.

In connection with application of roofing materials, you'll hear the word "coverage" frequently used. A roofing material designed for use with minimal lapping so that over most of the area of the roof there is but a single layer ... that is referred to as "single coverage." Most asphalt strip shingles are about a foot in width and intended for overlapping by 7 inches leaving 5 inches of "exposure." Exposure is the amount of a roofing product left exposed to the weather. With the strip shingles lapped just over half of their width, the roof will have two layers of shingles over most of the area and the coverage will be "double coverage."

The opposite of "exposure" is "weather lap." Both are distances measured up or down the slope of the roof. In the example above with a 5-inch exposure, the weather lap is 7 inches.

"Coursing" or "courses" refers to horizontal stretches of the roofing or siding. Continuing with the asphalt strip shingle example, the strips are applied so that their butts line up horizontally, and a single layer stretch of shingles is called a "course."

Certain parts of the roof will be referred to and these have become fairly well known, such as:

- ridge—the top edge where two sloping roof surfaces meet and divide the downward flow of rain water;
- valley—the downward sloping juncture off two roof surfaces that are at right angles with one another on an inside corner of the building;
- hip—the downward sloping ridge-like juncture of two roof surfaces on an outside corner of the building.

Flat roofs and those with very low slopes do not have the above shapes, and flat roofs require special types of roofing materials.

The lower parts of a roof are called the "eaves," the outward facing for which is the "fascia" and the underside of which is the "soffit." The metal water drainage channels hung on the fascia were once called "eaves troughs" but are now commonly re-

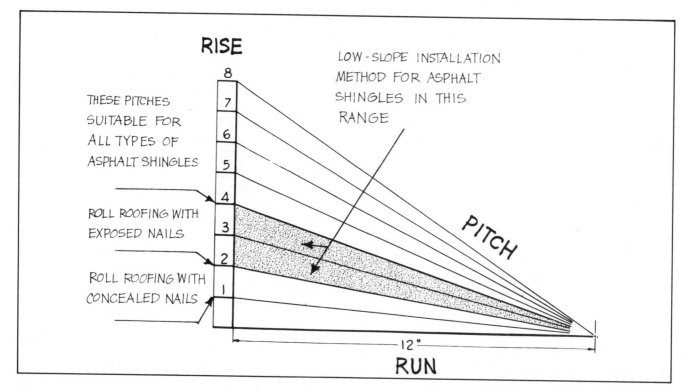

Roof slopes portrayed in this sketch indicate the suitability of asphalt shingles for various slope conditions. Rise, run, and pitch are common terms used when discussing roof slopes. And it is common practice to describe a slope or pitch in rise-to-run figures, such as "3-in-12."

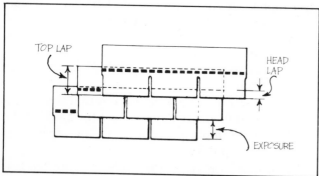

Asphalt shingle terminology includes the roofing terms indicated. Whether a homeowner contracts the work or does it himself, he should acquaint himself with the basic fundamentals of the work so that he can talk intelligently with contractors or suppliers. The drawings here come from the "Asphalt Shingle Installation Manual" put out by the Asphalt Roofing Manufacturers Assn.

ferred to simply as "roof gutters." At certain points, usually near the corners of the house, the gutters have outlets to which are connected "downspouts," once called "leaders."

There are other words applicable to roofing work and you'll be reading most of them in the next several chapters, but the foregoing will do for a starter.

This subject of understanding roofing terms and learning the basics could easily mean dollars-and-cents saved in dealing with re-roofing contractors or in discussing your needs with material suppliers. You won't become an expert overnight but by acquainting yourself with roofing information, you'll

be able to talk with contractors or suppliers intelligently, better judge their capabilities and knowledge, and thus be a more astute purchaser of surfaces or materials.

Re-roofing Complications

Quite obviously, some roofs are easier to re-roof than others. Steep roof slopes, for example, with a pitch or slope of 12 inches of rise per foot of run or horizontal distance, or greater (45° or more), usually involve following special instructions for application of the usual seal-tab asphalt strip shingles. The factory-applied adhesive spots work best for lower roof slopes, where the weight of the shingle plus the sun's warmth softens the adhesive and completes the bond. With steep slopes, the bonding process may be less effective and need additional spots of roof cement.

Steep roof slopes plus the presence of frequent roof dormers complicate re-roofing even more. More time is taken to fit shingles around dormer valleys and ridges. These are difficult areas to climb around, with poor footings increasing the hazard of accidental slips or falls. Steep slopes on your home should be a weighty factor in choosing to have a re-roofing contractor do your work rather than attempting to do it yourself. If the decision is to do the work yourself, acquire some first-hand knowledge in advance about working on steep slopes by observing experienced roofers doing their work. You can get job locations from roofing suppliers or by calling roofing contractors' offices. Observe the precautions taken by workmen. You'll also see some careless acts that you can guard against.

Perhaps the one most significant factor in re-roofing work is the condition of the existing roof sheathing underneath the roof covering. If the sheathing boards have, through the years, remained essentially dry throughout all areas of the roof, you are in luck. The probable need for replacing moisture-damaged sheathing is minimized. You can inspect the condition of the sheathing, of course, from below in the attic space. Look for uniformity in color, for easy splintering of dry wood, by picking with a screwdriver. Use the screwdriver to poke and dig in any discolored spots to verify the wood's firmness. Soft easy-to-penetrate wood spells trouble because the softness results from a moisture-related condition. Soft and nearly-soggy portions indicate roof sheathing boards that need replacement.

Lower sections in the roof are most likely roof areas for moisture to have wrought damage over a period of time, especially along the eaves. And these are areas difficult to inspect closely from the attic space, particularly on low-slope roofs. Do your screwdriver poking and probing from outside the house removing, if necessary, portions of the roof edging or fascia. If the wood sheathing, wood fascia, wood trim and rafter ends have been accumulating moisture and slowly rotting, the evidence will be more readily apparent than you might think, usually in the form of a general looseness of nails and easy lift-up or moving of the materials.

Rotting roof sheathing boards mean extensive repair work is likely in addition to just re-roofing. New roofing materials won't last long if they're applied over old rotting boards that won't hold nails. If the boards are damaged and soft in spots at different points on the roof, the recommended procedure is to remove the old roofing material completely, then remove the damaged sheathing boards and proceed with the new roofing.

If moisture-damaged boards appear to be limited to the lower eave slopes, it will not be necessary to remove all the old roofing; just the portions along the eave can be removed, boards replaced and new starter strips doubled to meet the old roofing material before applying re-roofing.

Another roof sheathing condition you may find is sagging that results in waves on your roof surface. This sagging between rafters is usually the result of simple aging, but it does mean that new roof sheath-ing will be required, and in most cases it can be nailed up right over the old wavy sheathing boards using longer nails to penetrate framing members at least 1¼ or 1½ inches.

For old wood-shingled roofs, the same probing for moisture damage of the shingle should be carried out. Visual evidence of such moisture may be present in the form of a shingle cupping or edge-curling. Shingles that give any evidence of decay need removal. If prevalent over many parts of the roof, entire removal is advised.

Roof Modifications

Many homeowners, at the time that re-roofing work is needed, begin to wonder about the feasibility of making certain changes in connection with the roof. One item commonly considered is the addition of roof skylights to bring daylight into an attic room or stairway. Skylights are a very easy-to-accomplish addition to handle when re-roofing. Skylights themselves can range from just a simple flat or dome plastic glass unit, to a complete aluminum-framed unit that operates like a window to provide ventilation as well as daylight.

Dormers Two types of dormer construction are commonly used. A window dormer is usually just a little wider than necessary to accommodate a window in the dormer face. The principal purpose is to give daylight and ventilation to an attic room. A shed-type dormer is wider, may contain several windows in its broad face, and the purpose is usually to provide more headroom for the attic living space.

With either dormer, re-roofing is a good time to consider dormer additions. The framing work is easily accomplished and can be done from within on the attic floor. Two factors here, however: (1) re-roofing contractors are not generally qualified as remodelers and the dormer work must be done in advance of re-roofing by a carpenter contractor, home improvement firm, or yourself; (2) dormers constitute a change in the home's structure and therefore work should not be started without obtaining a building permit.

Eave extensions Many lower-cost homes were built originally without any roof overhangs. Rafters were cut off right at the edge of the exterior wall and a frieze or fascia board applied to rafter ends. The lack of roof overhang often causes gutter overflow water to run down the face of the home's exterior siding and result in stains or discolored spots, and possible damage to siding edges and trim around windows and doors.

The lack of an overhang is something that can easily be corrected at the time of the home's re-roofing work. Removal of the fascia or frieze board allows access to the rafter ends, and rafter extension pieces of 2 x 4 or 2 x 6 can easily be nailed to the sides of the existing rafters. New sheathing is nailed to the rafter extension tops and the re-shingling work begun at the lower roof edge. A similar modification to extend the rake slopes on gable ends is also reasonably simple. On most houses, adding roof overhangs at eaves and rakes improves the home's appearance and eliminates a "boxy" look.

Other style changes In recent years, as new home prices continue to leap upward, increasing numbers of families have remained in older homes but have strived to achieve a more up-to-date appearance. They have used other roof changes which, though fairly simple in respect to the construction work involved, can bring about a surprising change in a home's appearance.

One such modification involves remodeling the gable-end roof peaks into a roof hip. Not a full hip in which the end slope continues down to eave level, but a partial hip that sort of slices off the peak at about window-head level. In other words, the new hipped portion comes down to near window-head height. A similar easing of the ridge peak on dormers brings about an entirely new roof style although the modification work is limited to just a few ridge peaks.

A more distinctive change in home appearance, but one involving considerably more re-framing work, is the addition of mansard roof framing to the top portion of exterior walls. Mansard roofs have become extremely popular and are used quite extensively in giving stores and commercial buildings a new appearance. The same process will work readily with certain kinds of older homes.

There are basically two types of mansard roof framing, one suitable for use with one-story construction, the other appropriate for two-story homes. The pre-existing condition that allows easy conversion to the mansard roof style is the existence of a flat or shallow-slope hipped roof. It is particularly easy to add the mansard slope to the hipped roof. Soffit materials, if any exist, are removed. New steep-slope mansard rafters are nailed at their tops to the existing rafters and lookout framing provided at their base. Sheathing is added if conventional materials are to be applied, but some types of man-

sard roof panels can be applied to wood strips rather than completely sheathed mansard rafters.

On one-story mansard roof additions, the vertical slope is shallow and comes down only to window-head height. On two-story mansard additions, the steep rafters are carried down to floor level of the second floor. In other words, the exterior siding above the first floor is now part of the mansard roof with appropriately framed openings for second floor windows. It's safe to predict that this type of home improvement will become commonplace in the next decade.

Roof-Related Jobs

The discussion in this chapter has dealt with the extra work that may either be necessary or desirable because of convenience in doing it at the time a re-roofing job is being considered.

There are certain other roof-related jobs that you may have to consider or may desire to do at the time of re-roofing or shortly thereafter. The common renewal jobs are, in the order of their frequency: replacement of gutters and downspouts, replacement or recovering of fascia and soffit materials, and the provision for more or adequate ventilation to attics and other under-roof spaces. These renewal materials are supplementary items for most siding manufacturers, and the re-siding specialists are the tradesmen most often applying them.

At this point, you may be convinced of your own capability in doing the re-roofing and associated work, or you may just be toying with the idea that you could do it yourself. The suggestion is that you follow the same procedure as that ordinarily followed by homeowners who know from the start that they will not do the work themselves. After you've visited some supply houses such as retail lumber or home center outlets furnishing roofing materials and after you've called in several re-roofing contractors or home improvement firms, then make your do-it-yourself decision. You will at that point have learned or confirmed the complications that may be involved, and you will have a much closer idea of what money you can save by taking on the task yourself.

About Roofing Contractors

Homeowners usually reach roofing contractors via the phone book's Yellow Pages. The normal reaction of homeowners when they check the classified category under "Roofing" is: "I didn't realize there were so many!"

This reaction also often indicates a wonderment about the several different kinds of roofing businesses, some apparently specializing in certain kinds of work and others being broad-ranged and, from their ads, being capable of handling almost any kind of roofing job.

Types of Contractors

As a general rule, more specialists in roofing will be found in metropolitan areas surrounding larger cities—and more roofers in smaller towns who do all kinds of roofing work. However, in recent years, many smaller towns and rural areas are being served by a few specializing roof contracting businesses based in some not-too-distant population center and doing business over a multicounty area.

Since the intent is to help the reader be more selective, this list includes those types of firms you should not contact as well as those you should.

Flat-roof specialists Their advertisements indicate emphasis on commercial-industrial roofing work, with main interest in the application of hot roofing products, the type of installation known either as "tar-and-gravel" roofing or "built-up roofing." Even if a small portion of your home's roof is flat with a tar-and-gravel surface, the big specialist operator will not be much interested in doing such a small job, unless it is at a big price.

Old establishments The roofing business "founded in 1896," now in its third generation, can do quality re-roofing work, but it has also learned about the hazards to profit-making and ways to allow for them. And it has learned to strongly recommend the extra work in repairs or supplementary installations and make the most out of a re-roofing job.

Roofing and insulation Usually this combination of services relates to flat or low-slope roofing applications on new commercial buildings where the subcontracts cover the installation of simulative board materials as well as built-up hot asphalt roofing materials.

Roofing specialists "We specialize in residential work ... all kinds of re-roofing and repairing."

Be alert for this kind of advertising message; it indicates the roofer knows what he should about residential roofing, and he recognizes the frequent need for minor roof repairing as well as re-covering.

Roof shinglers Some roofers specialize in doing only shingle work, asphalt or wood shingles; some shinglers will work only on new buildings, and many work only as subcontractors to home or apartment builders. If your home is to be re-shingled, you'll do well in both workmanship and price by calling in a shingle specialist who does re-roofing work.

Spray applicators Re-roofing firms which are mostly found in medium-sized or larger metro areas with many industrial buildings. They may use either a hot or cold process spray method, usually formulated for application on old built-up roofing. Despite claims made by spray applicators or patented spray manufacturers, be very wary about the use of such materials for re-coating residential roofs.

Roofing and construction Re-roofers skilled in carpentry trade will usually offer to do supplementary or associated work in conjunction with re-roofing, such as eave repairs, siding, roof decking. Some companies do both roofing and siding work though not often on the same home under one contract. The one-man jack-of-all-trades works under this banner, and conscientious men can be found, but most homeowners desiring a contractor probably want someone with more specific knowledge about re-roofing problems.

Slate and tile roofers Not many specialists are left in this category because these types of roofs have diminished in popularity except for a few areas of the country. If your home's roof is slate or tile and needs repair, call in a re-roofer with experience in this kind of roofing. If you decide to remove the old slate or tile, call in a roofer experienced in the new roofing you want.

Trouble-shooters Some roofers consider themselves expert in solving the leak problems or damage problems without a complete re-roofing installation. If your preliminary inspection reveals only a few trouble spots, your best bet may be calling in such a specialist.

Bear in mind that the larger firms, having a number of work crews, are more likely to be interested in commerical-industrial work than in residential. There are exceptions; it depends on how busy they are.

Licensed—Bonded—Insured Workers

One or more of these three words often appears in the ads, the stationery, or the proposal forms of a roofing contractor; 99 times of a 100, the homeowner will be duly impressed but will not really know why.

These are, in the contractor's mental image of himself, words that are supposed to communicate to prospective customers his competence. The words "licensed," "bonded," and "insured" give little indication of the contractor's capabilities.

The words do have meanings that vary from one location to another, and one situation or contractor to another. In an indirect manner they say to a prospective customer that this contractor is perhaps a lesser risk to the homeowner than some contractors who may not be licensed, bonded, or insured.

In explanation of these terms, let's go in the reverse order, because to the individual homeowner with a re-roofing job, a contractor's insurance is more important than his bondability, which in turn has greater significance than the contractor's license status.

Insurance Insurance in home remodeling work relates to accidents. Roofing work is accompanied by slips, stumbles and falls. Sometimes by scaffolds that give way or weak roof decks that won't hold concentrated weights. Whatever the cause, the result is an injured workman.

In most states workmen's compensation laws provide for employer insurance covering such injuries and the owner of the property does not get involved. Trouble can start, however, with small contractors who don't carry insurance of any kind or, because of their business seasonality, have let their insurance lapse. As the owner of the property on which the accident occurred, the homeowner is ultimately responsible. He can, however, protect himself by having the contractor certify in his job proposal that the contractor carries appropriate liability or compensation insurance. The ordinary homeowner's insurance policy does provide coverage for certain kinds of accidents around the home and property, but most will not extend to accidents involving workmen engaged in repair or remodeling work.

Bonds The homeowner should understand that there are various forms of bonds and that all bonds are, in reality, a type of insurance. For contracting work there are bonds that cover payments and bonds that cover performance. A bonding company acts like a trustee. The bonded contractor pays the bonding company a fee that's comparable to an insurance premium. Then, in the case of a payment bond, if the contractor does not or is unable to pay material suppliers or subcontractors, the bonding company does. In the case of performance bonds, the bonding company will take over the job and arrange for some other contractor to complete it if the original contractor cannot for any reason.

The above describes typical bonds issued only to cover a specific construction project. They are issued, however, usually only for larger construction or remodeling jobs that involve contracts of about $50,000 or more.

Some contractors carry continuing bonds that are applicable to their regular smaller jobs of various kinds and it is this type of bond that a re-roofing firm is likely to have. From their customer's viewpoint, the fact that a contractor is bonded is more an assurance that the job will be completed than any indication of quality workmanship. However, bonding companies do investigate company records, and for some contractors as well as for some jobs, the cost of bonding may be prohibitively high. The fact that a re-roofing contractor is bonded does have some value ... the fact that a bonding company considers this firm likely to properly perform and complete its work.

In roofing work, there is another kind of bond that frequently is involved on larger roofing jobs, particularly those on commercial or industrial buildings—it relates to the performance of the completed roof for a certain number of years. Roofing material manufacturers often prepare specifications for such bonded roofs. The specifications may call for certain application procedures, a certain number of roofing material layers, and other installation details. When a roofing contractor complies with these specifications, a bonding company will then issue a bond on the roof guaranteeing the roof will stand up for a certain period of time without developing leaks or requiring maintenance.

A roofing contractor who is experienced in meeting bonding company requirements may or may not be a good contractor to handle re-roofing work on a home. Many such contractors, of course, are not interested in doing residential re-roofing work. Of those that are, their experience will have provided them with knowledge of first-class roofing methods for the type of work being done under bonds. But re-roofing of homes is a different kind of work. And the contractor who knows the best methods is also the one who knows the shortcuts that can make up for the lower prices that may be quoted on residential work.

Licensing

Contractor licensing may be on a statewide basis or it may be by a local municipality. Licensing is essentially part of the political process. The qualities that are looked at by a politician or bureaucrat in order to grant a man a license to sell liquor are not much different from the qualities looked at to grant an architect or a plumber a license: the formalities are looked at; the experience is noted. The absence of negatives seems to play a role. And the license is issued. The system favors those already in that facet of business. It favors the established way of doing things. But it does not provide a measure of the competence of existing practitioners, only that of newcomers.

The fact that a contractor holds a license of one type or another really has little meaning for a homeowner desiring a re-roofing or re-siding job, because the homeowner cannot without considerable difficulty obtain information about methods of issuing licenses, what qualifications are necessary, or how rigidly the license laws are enforced.

Contractor Size-Up

If your home has a relatively simple low-slope roof, it will not pose much of a problem in re-roofing for either a contractor or for yourself. But steeper, more complicated roofs are the ones where more care should be used in accepting a contractor's bid.

Your chances for getting satisfaction from the re-roofing job will be greater if you take the time to stretch out your preliminary contractor contacts, ask a lot of questions and size up the contractor. Here's how to go about it.

First, using various sources of contractors' names, make a list of five, six or seven contractors you plan to call. Your list may include two or three names from the Yellow Pages, perhaps one or two from the classified services section of the local newspaper and another name or two from building or roofing supply companies. Try to limit the names to businessmen who appear to specialize in residential re-roofing work, and in the application of the kind of materials you believe you want.

Next, plan to spend several evenings or Saturday mornings talking with the selected contractors. Try to separate appointment times sufficiently so one contractor won't come while another is there. But let each one know that you are talking to another contractor or two.

The better contractors will take their time, perhaps walk off house length and width and probably ask several questions as to what you have in mind. The better ones will anticipate some of the things you planned to ask about. They will investigate roofing or siding conditions much more closely.

One item about which you will probably have to question even the better contractors is time. Approximately when they will be able to begin work on your job and how long the work will take. Will bad weather cause job delays and what they will do about such delays? Most contractors like to hang a little loose on specific times, particularly with respect to starting times. They want a little flexibility when they are not yet sure how their working schedule will fit in with this new job.

After your group of contractors has been called in, one will either not show up or will fail to send in a proposal. A couple will fail to impress you sufficiently and perhaps another will have an unusually high bid. That will leave you with two, three or maybe four estimates and proposals to consider. The prices quoted will vary, sometimes substantially. But this variance won't mean much until you note each proposal and the scope of work that each indicates will be done. Some will indicate repairs of sheathing boards along house edges. Replacement of gutters will be included with one or two of the re-roofing quotes. It is your job now to sort out these proposals, to recall the impressions made by these various contractors—their sincerity to answer your questions and be of help to you.

You may still be undecided between two of the proposals and may wish to pursue two additional steps before fully making up your mind! First ask for customer references to inquire about a previous job; try to get names and phone numbers of two or three homeowners that have recently had similar roof work done by the contractor. Then ask for a current job similar to yours and where you can stop by to observe his crew in operation.

The latter course is suggested where a homeowner has a more complex roofing job and wants to see primarily the equipment being used and the order of procedure. Since roofing jobs vary greatly, it may not be possible for a contractor to provide this immediately because all of his current work could be of a different type. But it does no harm to try, and to take the time to see how his crews do behave. It gives you confidence to proceed when you see another, similar job done without hitches.

The last step before signing contracts with a contractor, or accepting his proposal form, is to verify that all the items you want are included. One such important provision could be with respect to payment for work. If you're depending upon a home improvement loan, include a provision to the effect that this agreement with the contractor hinges upon your receipt of the loan from such and such lending institution. You will, of course, have to notify the contractor when the loan has been approved so that he can then proceed.

Do not be careless about details in the contract or proposal. It is the sole method of formal communication from a legal standpoint. So make it read right before you sign it.

Choosing Re-Roofing Materials

Funny thing about asphalt shingles. Years ago, you could only get them in red, green, and blue. You can still walk into some building supply dealers and look around for, but not see, any shingle displays. Ask the clerk and he'll say: "Yeah, we got asphalt shingles: What do you want, red, green, or blue?"

These dealers haven't changed, but asphalt roofing has. And what you can choose from today in asphalt shingles is enough to make your head swim. Obtaining what you want in shingles is a little complex, but first you have to know what's available; then, you have to know who's selling it. And that's the aim of this chapter.

In the re-roofing of homes, asphalt-strip shingles are way ahead as homeowners' favorite choice. But in West Coast areas, wood shingles and shakes are extremely popular. In other parts of the country wood shingles and shakes are found in the re-roofing of moderate-to-higher income homes. As a general rule, asphalt strip shingles can be easily applied over old wood shingles, and vice versa.

In some southern and southwestern states, clay tile roofs have been popular in combination with Spanish and Mexican architecture. Their use has diminished greatly and has given way in part to new precast cement and steel tile units. Also popular hard roofing materials in the past were slate —for luxury homes—and asbestos shingles for economy homes. Both have almost disppeared from the residential market, although slate roofing applications continue on commercial and institutional buildings.

There are metal roofing materials, but they are not intended for residential roofing. Found mostly in small towns and rural areas, metal roofing sheets are used predominantly on shed, warehouse and farm buildings. One exception is an aluminum product called a "shingle" or "shingle-shake." These are heavily embossed to simulate the appearance of handsplit wood shakes. They are really panels of aluminum about a foot wide and 3 or 4 feet long with interlocking edges. Though relatively expensive, these shingle-shake panels have become popular with the mansard-type roof boom.

Where possible, avoid built-up gravel roofing. This hot-application procedure is ordinarily employed on flat roofs but saw some usage on low-slope residential roofs until a low-slope shingle application was developed. The minimum slope for normal asphalt strip shingles is 4 inches rise per foot of run.

However, most manufacturers of shingles suggest a special cement-down method for slopes between 2-in-12 and 4-in-12. If your home has a built-up tar-and-gravel roof, re-roofing may be done with shingles if the slope is suitable, but otherwise should involve a re-coating of hot-application asphalt and roofing felts by a qualified roofing contractor.

In built-up roofing applications, you will hear such terminology as "20-year roof." This refers to the number of layers of felts and asphalt and to the traditional practice with commercial or institutional buildings of providing the building owners with a performance bond in which the roof is guaranteed to endure for a certain time period. In residential applications, bonding is not commonly done, but in recent years many manufacturers have offered a warranty on the roofing materials.

The life of a roofing material depends upon a number of factors. The most unpredictable factors are apt to be of a localized nature ... weatherability in the particular location of the home, type of wear or exposure in that specific locality.

Under normal expectations, economy grade materials can be expected to perform reasonably well over some seven to ten years. Middle-of-the-road grades of materials will probably not begin to show evidence of deterioration until about fourteen or fifteen years ... twenty if you're lucky. Most better grade roofing lines should have a life expectancy of twenty-to-thirty years unless the home experiences exceptionally severe weather conditions, or the materials are used inappropriately.

A homeowner considering the purchase of roofing materials either for do-it-yourself application or for use by a contractor should keep one point in mind. The labor portion of the re-roofing cost remains just about the same whether good or poor quality materials are used. Thus, over a long term, the decided advantage in "cost per year" of a re-roofing goes to the better quality materials.

Asphalt Shingles

In most areas of the country, where asphalt shingles predominate in re-roofing material as well as on new homes, the asphalt strip shingle is by far the favorite among the several kinds of asphalt roofing. When strip shingles were first introduced, they were commonly referred to as "3-tab" shingles. The 1 x 3 foot strip had two short ½-inch cutouts at foot intervals to give it three shingle tabs. The 3-tab is still widely used today, but to this type have been added the no-tab, the irregular-butt shingles plus several types of double-layer laminated shingles, all still keeping the same approximate 1 x 3 foot dimensions.

The trend has been toward heavier shingle weights, and more colors and styles. For many years, the standard, widely used weight of shingle coverage was in the range of 210-to-240 pounds per hundred square feet. But in recent years new manufacturing methods, plus the double-layering and use of larger-heavier granules, has run shingle weight to 300 pounds and more. As in other fields, the better quality added features and longer durability bring higher prices.

Although the strip shingle long ago became the preferred type, it was just in the past decade that the "seal-tab" provision for wind resistance became commonly used. This feature incorporates into the shingle a row of adhesive spots at proper positions. Under the heat of the sun the spots soften, and the tabs of the next course above settle down in the soft adhesive to give an effective protection against wind lifting the tabs.

The success of the seal-tab technique has diminished use of earlier individual or interlock wind-resistant shingles.

Normally, for standard size strip shingles, the overlap of one strip to the one below is designed to leave an exposure of 5 inches. This provides double coverage ... that is, every point on the roof has at least two layers of the basic roofing material.

Just a few years ago a new kind of asphalt shingle made its debut—the fiberglass shingle. This involved more than a mere cosmetic change; it meant an entire new manufacturing process. With rapid expansion of production facilities (for a relatively slow-moving industry) fiberglass shingles are now being marketed strongly and will most certainly become widely used.

The new shingles are made using a fiberglass mat as a base for the asphalt instead of the old organic felt base. The result is a significantly improved product. With the felt base the asphalt impregnates the base which acts as a "backbone," while the asphalt is really the waterproofing material. The mineral surface granules keep the asphalt-and-felt combination from drying out in the heat of the sun, and protect it from weathering.

With fiberglass shingles, the felt base is replaced by a glass fiber mat that is considerably stronger than the felt, although the glass fiber mat is less bulky. For this reason, shingles made with the fiber mat contain about 50 percent more asphalt for the same overall weight of shingle. This means better weather-resistance and longer life. Also, the fiberglass shingles are more fire resistant. They earn a Class A fire rating from Underwriters Laboratories, in contrast to the Class C rating, the highest that felt-based shingles can earn.

A further contrast between the two types is that fiberglass shingles cannot absorb moisture vapor. Accordingly, they do not shrink and swell with outdoor relative humidity ... changes that helped speed the wear and weathering of felt-based shingles. Fiberglass shingles are also lighter and easier to handle per strip, and the material cuts more easily.

Color facts It's the tiny ceramic granules that give asphalt shingles their color. And they have played a very big role in the greatly improved appearance of asphalt-shingled roofs in the past ten years.

The granule manufacturers for a long time were hampered by sticking with the old traditional shingle colors. Any variations were accomplished by blends of a basic color rather than any true move toward different colors. Then, along came residential air conditioning in the early 1960s and, particularly in southern states, the demand for white shingles.

Earth tones The change toward white made it easier for another color trend to develop just a few years ago—growth in the use of rustic or natural wood-toned exterior sidings. Soon, the asphalt roofing manufacturers were offering a range of brown, tan and beige colors ... colors incidentally that were visually more effective with the earth-toned color schemes that were becoming more frequent.

The idea of brown as a "better" home color appears to have scientific support. The German Lusher Color Test text indicates that brown is a color which is associated "with a place where people feel secure and where they enjoy the creature comforts ... a place where importance is placed on roots, on hearth and familial security."

At present, the appearance of asphalt shingles is further being diversified by the textured look. This is being done by some manufacturers through the double-layering process, by irregular butt lines, and by two-toning the shingle colors.

Roof color and texture are perhaps more important to the homeowner who is re-roofing than to the owner of a newly purchased home. It gives the owner of an older home another tool in individualizing his home. But certain fundamentals should be observed:

• choose a color that will fit your home's style but also one that complements your exterior siding or siding-to-be;

• pastel color shingles appear to fit modern or contemporary styling of a home while medium-range

earth colors fit homes in rural or wooded areas;
- bright-colored roofs can tie together older-styled homes that have many gables, wings or dormers;
- heavily textured roofing seems to be at its best with simple and straight roof shapes.

In considering colors for your roof (and walls), remember that white and light colors reflect heat, and darker colors absorb heat. So, in the South, where more energy may be expended for cooling than heating, try to select lighter colors; in the North, consider darker colors. But remember, even in the North summer can be very hot.

Shingle types Asphalt shingles are normally selected by their weight class and certain other features. The better quality shingles weigh more and cost more. Normal range for double-coverage shingles is about 210-240 pounds per square, and the double-layered and laminated shingles are 350-400 pounds per square.

Asphalt shingles of certain types carry certification labels of Underwriters Laboratories signifying their fire and wind resistance. A Class C label indicates shingles that have been UL-tested for light fire exposure and found that they will not readily ignite nor support the spread of fire. Nor will these shingles add to the hazard by the emission of burning particles that might ignite new areas.

A more fire-resistant asphalt shingle carries a Class A UL label. This means the shingle is even better than Class C in its fire resistance.

The label on UL-tested fire-resistant shingles may also carry a notation with respect to wind resistance. To have such a notation, the shingles must withstand test winds of 60 mph continuously for 2 hours without a single tab lifting. These wind-resistant shingles are the self-sealing types previously mentioned.

Some southern areas of the country are located where roofing shingles are subject to natural growths such as certain fungi or algae. The growths are usually accompanied by discolorations or stains. While the stains can be removed chemically, a permanent cure is more difficult. In recent years, a number of shingle producers have offered a white-colored line of shingles that are fungus-resistant, and thus have a better chance of remaining white without discoloration.

Textured shingles Tops in most shingle lines are the double-layered, laminated shingles designed to provide a heavy, textured appearance. Other texturing devices include two-toned colors, irregular butts and staggered tabs. These shingles are more individualistic. Though higher in price, they bring more durability and better appearance. Some lines employ embossing; some emphasize a thicker butt line.

In short, the innovations in the past few years mean a much greater selection is available. For example, the following lines of asphalt shingles are now available from one leading manufacturer:

A. 12 x 36-inch strip shingles
- 2-tab self-sealing fire-resistant Class A, 25-year warranty
- 3-tab self-sealing standard weight Class C, 15-year warranty; medium weight, 20-year warranty; heavy weight, 25-year warranty
- 3-tab fungus-resistant white, Class C, 15-year warranty

B. recent innovative additions
- 15 x 40-inch 2-ply laminate self-seal, 25-year warranty, shake projections
- 15 x 40-inch random-tab self-seal Class C, 25-year warranty
- random-tab self-seal Class C, 25-year warranty; less fire-resistant, Class C, 18-year warranty.

All of the latter and two of the square-tab types also have matching-color preformed 12-inch hip and ridge shingle available to simplify installation and improve appearance.

Availablity Now that you're delighted to discover so much to choose from, prepare for a little deflation. Although the many types described above are being made, you're not going to be able (at least for a while) to run down to the nearest building supply store with your van to pick up a few bundles.

The reason is that the retailer is stocking less and the more numerous and strategically located wholesale distributors are stocking more. As a result, small contractors would like to have you choose your roofing shingles from the limited choices that may be carried by their nearest supply sources, often a retailer that offers a contractor discount. When the homeowner wants a broader range of choice, the contractor has to contact the roofing distributor. On the newer materials, he has to obtain samples and prices before he can quote you accurately. And he may not have established credit at the distributor which means paying cash. On top of all this, he doesn't know delivery times and how soon before he's ready to do your job that he can order out the material. So, many contractors discourage homeowners who try to choose materials other than those the contractor normally uses.

For the do-it-yourself homeowner, the availability of wide choice in roofing materials is also complicated by the limited stocks of retail dealers. And while dealers or home centers are happy to order non-stocked materials, they seldom have descriptive literature or samples covering even all of one manufacturer's line, let alone those of other manufacturers. For the homeowner who feels limited by

Recommended Shingle Exposures

PITCH	NO. 1 BLUE LABEL			NO. 2 RED LABEL			NO. 3 BLACK LABEL		
	16″	18″	24″	16″	18″	24″	16″	18″	24″
3 IN 12 TO 4 IN 12	3¾″	4¼″	5¾″	3½″	4″	5½″	3″	3½″	5″
4 IN 12 AND STEEPER	5″	5½″	7½″	4″	4½″	6½″	3½″	4″	5½″

Maximum exposure recommended for roofs:

Shingle Coverage per Bundle

Approximate coverage of one square (4 bundles) of shingles based on following weather exposures

LENGTH AND THICKNESS	3½″	4″	4½″	5″	5½″	6″	6½″	7″	7½″	8″	8½″	9″	9½″	10″	10½″	11″	11½″	12″	12½″	13″	13½″	14″	14½″	15″	15½″	16″
16″ x 5/2″	70	80	90	100*	110	120	130	140	150‡	160	170	180	190	200	210	220	230	240†	....	....	....	....	....	....	....	....
18″ x 5/2¼″	....	72½	81½	90½	100*	109	118	127	136	145½	154½‡	163½	172½	181½	191	200	209	218	227	236	245½	254½‡	....	....	....	....
24″ x 4/2″	....	....	....	....	....	80	86½	93	100*	106½	113	120	126½	133	140	146½	153‡	160	166½	173	180	186½	193	200	206½	213†

NOTES: * Maximum exposure recommended for roofs.   ‡ Maximum exposure recommended for single-coursing No. 1 and No. 2 grades on sidewalls.
† Maximum exposure recommended for double-coursing No. 1 grades on sidewalls.

Coverage for Handsplit Shakes

SHAKE TYPE, LENGTH AND THICKNESS	Approximate coverage (in sq. ft.) of one square, when shakes are applied with ½″ spacing, at following weather exposures, in inches (h):					
	5½	7½	8½	10	11½	16
18″ x ½″ Handsplit-and-Resawn Mediums (a)	55(b)	75(c)	85(d)	100	....	....
18″ x ¾″ Handsplit-and-Resawn Heavies (a)	55(b)	75(c)	85(d)	100	....	....
24″ x ⅜″ Handsplit	....	75(e)	85	100(f)	115(d)	....
24″ x ½″ Handsplit-and-Resawn Mediums	....	75(b)	85	100(c)	115(d)	....
24″ x ¾″ Handsplit-and-Resawn Heavies	....	75(b)	85	100(c)	115(d)	....
24″ x ½″ Tapersplit	....	75(b)	85	100(c)	115(d)	....
18″ x ⅜″ True-Edge Straight-Split	....	....	....	....	....	112(g)
18″ x ⅜″ Straight-Split	65(b)	90	100(d)	....	....	....
24″ x ⅜″ Straight-Split	....	75(b)	85	100	115(d)	....
15″ Starter-Finish Course	Use supplementary with shakes applied not over 10″ weather exposure.					

(a) 5 bundles will cover 100 sq. ft. roof area when used as starter-finish course at 10″ weather exposure; 6 bundles will cover 100 sq. ft. wall area at 8½″ exposure; 7 bundles will cover 100 sq. ft. roof area at 7½″ weather exposure; see footnote (h).

(b) Maximum recommended weather exposure for 3-ply roof construction.

(c) Maximum recommended weather exposure for 2-ply roof construction.

(d) Maximum recommended weather exposure for single-coursed wall construction.

(e) Maximum recommended weather exposure for application on roof pitches between 4-in-12 and 8-in-12.

(f) Maximum recommended weather exposure for application on roof pitches of 8-in-12 and steeper.

(g) Maximum recommended weather exposure for double-coursed wall construction.

(h) All coverage based on ½″ spacing between shakes.

what nearby retailers or home centers offer, inquire about locations of wholesale roofing supply warehouses. Go to such a warehouse or to the offices of a wholesale distributor, cash or check in hand, and you will find the wider choice of materials and a willingness to do business.

Roll roofing There are two types of this lowest-priced of all roofing ... smooth-surface and mineral surface. Neither type has ever been used much in modern housing for urban-suburban roofs. But it doesn't take much traveling or observing to see that before the early 1900s, roll roofing was a very common residential roofing material. And you can also observe among older houses that this roofing material was applied with laps parallel to the roof ridge, or at right angles to it.

Roll roofing continues to be manufactured and distributed, although most retail outlets today are in smaller towns and in rural areas. In these areas it is no longer used for roofing on prime homes, but often for cottages or seasonal homes. And it is still widely used for roof covering on outbuildings.

Roll roofing comes 36 inches wide and 36 feet in length, covering a half a square per roll with a normal exposure of 17 inches. The mineral surface is slightly higher in price than the smooth; the mineral granules used are very similar to, and give the same coloring as, asphalt shingles. Roll roofing is quite easy to apply and, with proper application, can give a very serviceable roof.

Wood Shingles and Shakes

First, the difference between them: shakes are split, while shingles are sawed; shingles have a relatively smooth surface while shakes have a textured surface due to the natural grain split; handsplit shakes have a heavier texture and thicker butts. Wood shingles and shakes are excellent materials for use as exterior sidings as well as for roofs.

Most shingles sold throughout the country today are of Western Red Cedar, although in some northern areas there are shingles available milled from local cedar species. However, Western Red Cedar has virtually become the standard of the wood shingle-shake industry. Cedar is a species, incidentally, whose cellular composition provides thermal insulative value, an unusual characteristic in roofings.

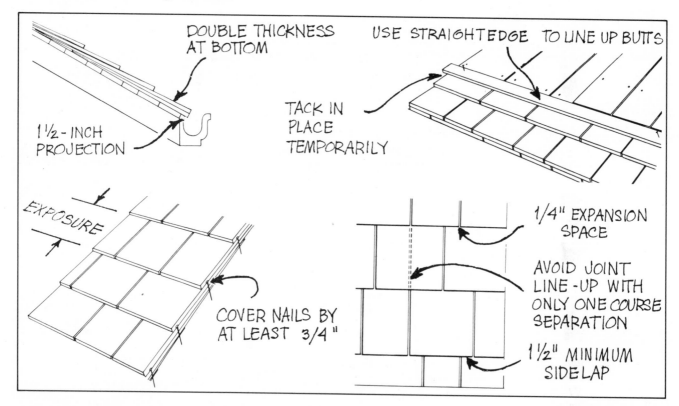

Other general characteristics of cedar shingles are good resilience plus excellent resistance to wind and natural elements. For roofing purposes, cedar shingles and shakes are furnished untreated and unfinished. The natural wood, left to weather, turns from its original reddish tan color to an attractive natural gray. However, homeowners interested in this color aspect should be forewarned that the change takes a considerable period of time and doesn't occur uniformly on the various roof slopes of the normal home. The natural graying can be assisted by a bleaching procedure.

Types and grades Roofing shingles are available in 16, 18, and 24-inch lengths and bundled four bundles per square of coverage. Grades include the following: premium No. 1 all-clear and edge-grain; No. 2 is mostly clear with flat grain and limited sapwood; No. 3 is a utility grade for economy applications.

Roofing shakes are available in 18 and 24-inch lengths and bundled five bundles per square of coverage. Grades: No. 1 handsplit and resawn; No. 1 tapersplit; and No. 1 straightsplit. Tapersplit refers to reversing the splitting block with each split in order to obtain a wedge-shaped shake while the block remains in uniform position for straightsplit uniform thickness shakes.

If your cedar shingle suppliers stock shingles and shakes that originate in western mills who are members of the Red Cedar Shingle & Handsplit Shake Bureau (RCSHSB), there are certain grade names and terms this association uses. The names "Certigrade" for shingles and "Certigroove" for standard shakes identify the graded shingle product, and "Certisplit" is used for graded handsplit shakes. The handsplits are thicker-butted, running 1¼ to 1½ inches while the standard shake butts are about ¾ inch.

Where non-western cedar shingles are being sold, the nonlabelled products will generally be offered in only two grades—No. 1 clear and utility grade—the latter having knots. In Western Red Cedar, the grade names are No. 1 Blue Label, No. 2 Red Label, and No. 3 Black Label.

Shingle usage When applied to roofs, the amount of exposure used with wood shingles varies with the degree of roof slope. Here are the normal exposures:

No. 1 Grade Roof Shingles

Slope of Roof	16 in.	18 in.	24 in.
4-in-12 or steeper	5 in.	5½ in.	7½ in.
3-in-12 to 4-in-12	3¾ in.	4¼ in.	5¾ in.

To estimate shingle needs for your roof, you will need fairly accurate house dimensions and degree of slope. You also need the approximate length of valleys and hips, since these take about a square extra for every hundred lineal feet.

Cost and other factors While there's a high desire and demand among homeowners for the appearance that wood shingles or shakes provide, the cost is sometimes prohibitive. Cedar shingles do impart a warmth and texture to the home unmatched by other materials. Yet, the facts of life are

Procedures shown in the above series of sketches are those suggested by the Red Cedar Shingle & Handsplit Shake Bureau: (1) cut away lower course of old shingles; (2) also cut back six inches from rake or gable edges; (3) nail wood strips along edges; (4) replace old ridge shingles with beveled siding strips, thin edge down; (5) add wood strip in shingles over the valleys to separate old metal from new; (6) apply new cedar shingles over the old ones using 5-penny rust-resistant nails. Left, in sketch form, are additional details to be observed when applying wood shingles. Wood shingle application is a relatively simple task, one within the capabilities of many do-it-yourself homeowners.

that wood shingles may run three times, wood shakes five times, the cost of standard asphalt strip shingles.

With respect to the durability or length of life, and thus cost per year, the two products appear to be roughly comparable, particularly with the newer figerglass shingles. Normal expectations for wood shingle life will be in excess of 20 years if the locale is one of limited roof moisture, and if the development of moss or fungi is not encouraged.

For re-roofing and re-siding purposes, cedar shingles and shakes are considered excellent do-it-yourself products—not at all difficult to apply and sometimes fun to do. They can be applied readily over old asphalt-shingled or wood-shingled roofs without removal of the old roofing. Old shake roofs will probably have to be removed to accommodate new roofing because the shake butt lines are thick and irregular.

You may wonder about wood shingle or shake roofs as a potential fire hazard. This writer knows of no statistical studies concerning fire hazard qualities of wood shingles. However, there is a new development in this respect: some producers of cedar shingles and shakes are now offering them with a fire-retardant treatment that earns the shingles a Class C fire rating by Underwriters Laboratories. And if these treated shingles are applied with a special plastic-coated steel foil underlayment, the resulting installation will earn a Class B fire-resistance rating.

Availability Wood shingles and shakes are more widely used in western states and thus westerners have more choice in the number of dealers, home centers, and other suppliers that carry good stocks. In the Middle West and East, suppliers are apt to stock only the No. 1 grade and, depending upon the wholesaler location, may or may not offer to special-order the type or grade you wish to have. Gradually, in recent years as the demand for top quality wood shingles and shakes has expanded, the limited availability situation is easing in many states east of the Rockies.

Other Roofing Materials

A few has-beens, a newcomer or two, some easy to apply and others difficult ... that is a one-sentence overview of the minority representatives in the roofing material picture.

While there may be a producer or two who will argue the following point, it is evident from declining usage that roofing products which at one time were commonly used on both the highest and the lowest cost homes are now being used on almost no homes, new or old. Slate roofs were a luxury home material in the 1920s and 1930s. Then, for a couple decades, asbestos shingles became widely used on economy house roofs and for re-siding of older homes. Today, slate roofing is used on institutional buildings and little else. Asbestos products have acquired a taint from the lung disease hazards that asbestos workers have been exposed to. While a manufacturer or two may still be producing what might be referred to as an improved form of what used to be called asbestos shingles, the product sees very limited application in either new homes or for re-roofing. Slate and asbestos shingles, though having shown a satisfactory performance record, were susceptible to breakage and chipping.

Among the higher-priced roofing materials are aluminum shingle-shakes. Use is made of the hyphenated word because that is the name the producers use for these preformed, embossed panels whose irregularly grooved or textured surfaces have been molded to simulate the coarse surface of handsplit wood shakes.

The aluminum panels come in various sizes: 12 x 36, 12 x 48, and 12 x 60 inches depending upon whose product is being used. The principal manufacturers now making and marketing these panels are the major aluminum companies: Alcoa, Reynolds, and Kaiser. The shingle-shake panels have their edges formed so that during application a four-way interlock of shingle-to-adjacent shingles is provided, making them impossible for wind to lift.

Aluminum shingle-shake panels have been utilized more on commercial buildings than residential. And a considerable portion of their increased usage is because of the boom in commercial remodeling work which makes use of mansardlike roof attachments. Aluminum shingle panels can be easily applied over a relatively light wood framework for such mansardlike exteriors.

Another type of metal roofing panel is called "Decramastic" roof tile. Made of relatively lightweight galvanized steel, covered with mastic and mineral granules, it also is finding widespread acceptance where the appearance of tiled roofs is desirable or as a material for use on the above-mentioned mansardlike exteriors. Available in one size with a full range of accessories, the 14½ x 34¼-inch four-tile panels are produced by Automated Building Components Inc. The product comes in a choice of nine colors and a weight of approximately 170 pounds per square.

Tiled roofs are considerably more popular on homes in milder climates. And while old, damaged or deteriorated tile roofing needs to be fully removed before any re-roofing occurs, the simulated tile mentioned above can be applied over old roofing using 2 x 2 batten strips.

Clay tile roofing goes back four or five thousand years to when it was a roofing material in China. In modern times, clay tiles have been used extensively on homes in the southwest part of this country and are still in limited use on commercial structures.

But another type of tile roof has been developing and spreading . . . concrete tiles. Sometimes referred to erroneously as "cement" tiles, these units had their origin during the last century in European countries. In the U.S., concrete tile roofing has been pioneered by the Monier-Raymond Co. whose product has been based on a patented concrete extrusion process using automated machinery that turns out a consistently high-quality product. These tiles are perhaps as durable, or more so, than any other type of roofing since they are sold with warranties running to as much as fifty years. The tiles are fireproof, immune to rotting and termite or rodent damage. They resist weathering of all types and actually grow stronger with age.

Said to be as much as 40 percent faster in application than clay tiles, the concrete tiles are applied over parallel 1 x 4 boards nailed to rafters (new construction) or to old roofing. The boards are applied horizontally (parallel to roof slope) and spaced about 12 to 13½ inches in order to take the 13 x 16½-inch tile units. These "Monray" tiles involve an interlocking side-lap of about 1¼ inches. The company recommends that headlap distance vary with home location: two inches in the Southwest; three inches in the rest of the country; four inches for high wind conditions.

Very definitely, a homeowner considering re-roofing with concrete tiles should give careful consideration to his home's roof structure and its load-carrying capability. Concrete roofing tiles weigh from 840 to 990 pounds per square, depending upon the amount of headlap. This is more than triple the weight of standard 240-pound asphalt shingles. The tiles may have an even greater weight when wet.

Copper sheet roofing is sometimes applied on architectural buildings such as churches and college buildings, shopping centers, stores and business offices. The common method of application is called the "standing seam" method. In recent years, use of another metal called "terne," a tin-lead alloy that bends and crimps easily, has increased on these commercial and institutional buildings. Neither type of roofing is considered suitable for normal residential re-roofing applications, even if homeowners could afford it.

Roofing It Yourself

Many can do it. It's not a difficult job, but it is a tedious one and often a hot, sweaty one. But you stand to save nearly half the cost compared to what a roofing contractor will quote you if he uses about the same grade roofing materials,

More than capability may be involved. Re-roofing can be very time-consuming. There may be preparatory work involving roof board replacement, or new flashings that must be obtained and fitted. Not to mention roof gutter replacement, the most common supplemental job.

Unless your home is a simple one-story house of 1,200 square feet of floor area or less, do not contemplate doing it alone. At least a helper, to help hoist shingles and do other assisting jobs, is needed. Better still for the do-it-yourselfer is a full-fledged neighbor or relative assistant who can do shingle nailing, plus a helper to keep both applicators supplied with their materials.

Low-slope roofs of 5-in-12 slope or less, and roofs without complications by dormers or wings, are easiest to work on. Mostly for reasons of safety, here are certain types of houses on which do-it-yourselfers will do themselves a favor if they contract out the work:

- roofs whose slopes are of 10-in-12 slope or steeper
- relatively steep roofs having numerous dormers, chimneys, wings, porches and varying slopes
- roofs on any homes higher than two stories
- any home having a substantial roof area of built-up tar-and-gravel roofing

This latter tar-and-gravel roofing sometimes occurs in the areas where the roof structure is relatively flat, such as over porches or the top portion of a mansard roof. The application of hot asphalt and roofing felts to provide a new layering over an old built-up roof section is work for the roofing expert. Don't try it yourself. If reconditioning of this portion of your roof is necessary, call in a contractor and have it done before you proceed with roofing shingles on the sloped portions of the roof.

The decision to allow old roofing to remain in place is one you'll have to make following a conscientious inspection. The principal factor affecting this decision will be the condition of roof sheathing boards under the old roofing. If the boards are in good condition without evidence of moisture damage, probably the best decision is to cover rather than remove. A possible exception might be made for a rather common situation already mentioned: where the lower eave areas have suffered moisture damage due to ice formation or damming during winter months.

In such cases, it is quite practical to remove the old roofing in a strip about 2- to 4-feet wide along the roof eaves. Any damaged sheathing boards in this area can be replaced, a couple layers of building felts cemented down, and shingle re-coursing that comes up to the old shingles can be applied before proceeding with the re-roofing work.

Some similar modifications may be necessary in preparation for re-roofing of wood shingle roofs. If making a change from the old wood shingles to new asphalt shingles, a close look for old loose or beginning-to-decay shingles should be made. Pull out any shingles in poor condition, remove nails and slip in replacement shingles. The work in replacing shingles needn't be perfect, but try to match shingle thicknesses to achieve relatively even surfaces. If shingle butts are fairly thick (over ½ inch), wood wedge-shaped strips ripped from 1 x 3s or 1 x 4s should be nailed butt-to-butt with the old shingles to furnish a broader support base.

If old wood shingles are so bad that the entire shingle covering ought to be removed before re-roofing, you may have to do some added work on the sheathing. With wood shingles on new construction, it is common practice to space out sheathing boards leaving 1½- to 2-inch gaps between boards. This is sometimes called slat sheathing. Where possible, it's desirable to allow old shingles to remain in place if the gapped sheathing has been used. Otherwise, if the re-roofing is to be asphalt shingles, the gaps have to be filled to become solid sheathing. You can have 1 x 6s or 1 x 8s ripped to fit the gaps; a uniform tight fit isn't necessary, but a relatively smooth surface should result. Sometimes the easiest and quickest way to obtain this smooth surface over old roofing or old roof sheathing boards is to simply add a completely new layer of plywood sheathing.

The general rule about the need for roofing felt or 15 pound asphalt breathing-type building paper is fairly simple: no felt or paper if re-roofing is applied over old roofing; otherwise, apply it lapping by half to obtain double coverage. This is easy because of the white lines provided by manufacturers for lap guidance. Because roofing felts or felt building paper (it has different names in different parts of the country but the product is essentially the same) are quite inexpensive, the paper offers cheap insur-

Before re-roofing, the base upon which new shingles are to be placed needs attention. Nail down or remove protruding roofing nails. Renail split shingles. Replace broken or missing ones. Whether the old roofing is asphalt or wood shingles, get a smooth firm foundation for the new roofing material.

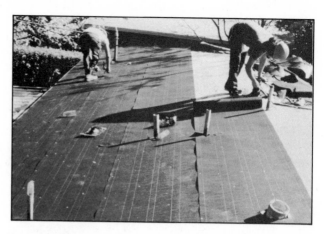

Roofing felt application is needed on pitches below 4/12, when old roofing is removed. Again, remove old nails or drive them down before applying the felt strips. Use care in alignment of strips so that the white shingling guidelines are kept parallel with the eaves and ridge lines.

ance as a suitable under-layer for asphalt shingles. The one exception to the general rule on felt requirement is: if old roofing shows any evidence of having a sustained moisture condition, remove the old roofing completely, including the felts, and start afresh. Moisture conditions may be evident in shingle cut-outs or spaces between shingles where there is a surface mossiness, fungus growth or oxidation discoloring that, under probing, reveals a slightly soggy area below the surface.

Tools & Equipment Needed

On any kind of job, the experienced do-it-yourselfer knows that it pays to anticipate just what tools, devices, and working equipment will be needed. It's not only a waste of time and effort trying to do a job without the proper tools, but it results in a sloppy job.

The foregoing discussion of reaching a decision on the removal of old roofing will add a few tools to your list. A prime tool is the medium-sized (24 to 30 inches long) wrecking bar or crowbar. This device is much easier for prying and nail-pulling than a claw hammer. With asphalt roofing a "square shovel" is fast and efficient in removing layers of shingles and felts. This type of shovel has a square rather than long handle, and has a more rigid edge for prying than does a scoop or coal shovel.

Generally, any effort to save the felt underlayment when ripping off old roofing will fail. The underlayment may stay in place but it's very apt to be damaged by cuts and breaks.

With old wood shingles, the removal work may go fastest using the wrecking bar to loosen the shingles and the shovel to take them up. The shingles will probably split quite easily during removal,

especially at points where they are nailed. The result is a bunch of roofing nails left in the sheathing boards with projecting nailheads. These must be pulled or driven down flush. Your best tools for this are a 12-ounce claw hammer and a 15-year-old son or daughter.

Nine times out of ten, the headachy part of re-roofing comes with the roof edges. Take care in your wrecking and removal work along roof edges not to break off sheathing boards or damage roof edging material. Part of the effectiveness of roof edging metal is its uniform support of shingle edges, and since part of the edging is visible from the ground, you don't want your tools putting dents in it or otherwise damaging its regularity or alignment.

Now, for re-roofing application tools. The claw hammer may prove better for your use in nailing down asphalt shingles than the traditional roofer's hatchet. This type of hatchet, with a large-area corrugated hammering head opposite the cutting edge, was originally the tool of the wood shingler because his work involved a lot of cutting of shingles to suitable width and this could be done at a single stroke with the hatchet. This is still a prime tool for applying wood shingles. Further details on the use of the hatchet and other wood shingling devices is provided later in this chapter.

With hammers and hatchet go nails. Roofing nails are hot-galvanized nails made of steel in 11 or 12 gauge with a ⅜-inch diameter head. With asphalt strip shingles, the quantity of nails needed is about 2½ pounds per square of roofing. For new construction 1¼ or 1½ inch long nails are commonly used. For re-roofing, longer nails should be obtained, their length depending upon the thickness of the old roofing material and the sheathing. With old as-

Cedar or redwood shakes are very popular today, both as siding and roofing material. When allowed to weather naturally these shakes help the home to blend in with the surrounding scenery. Photo courtesy of Vermont Weatherboard

This older home was remodeled with steel siding. Metal siding is used very often on older homes.

A home re-roofed with textured asphalt shingles in a red-brown-white blend, which complements the brick and trim.

A 1½-story contemporary style home enlarged with a garage-basement and family room above, plus bleached plywood siding and light brown shingles.

A close-up view of a cedar shake mansard soon after application.

Wood trim tends to be neglected on maintenance-free stone homes. Regular maintenance is vital to prevent wood trim from rotting.

This well-maintained traditional style stone veneer home has a fresh appearance with the choice of layered, irregular-butt shingles in a brown blend.

This two-story home is having its exterior modernized, with a mansard roof added to the second-floor areas by simple triangular 2 x 4 frames followed by batten strips and shakes.

Stained finishes for plywood are particularly appealing in the darker tones, as with the grooved reverse board and batten of pine plywood. Photo courtesy of Georgia Pacific

These dark wooden planks require little maintenance and will last indefinitely. Photo courtesy of Vermont Weatherboard

Clapboard is a very popular style and is common on older homes. Replacement is very expensive, and, if at all possible, the original siding should be repaired.

Plywood strip panels on this two-story home are designed for paint finishes. Their special resin-treated surface produces what is called "Medium Density Overlaid" plywood. The surface is uniformly smooth without any graining.

Aluminum and steel sidings come with smooth factory prefinishes as shown by the double 4-inch clapboard style on this home.

These two colonial homes show an excellent blend of traditional coloring. Notice the beautiful use of shutters and entrance treatment.

A proper blend of exterior materials will result in a beautiful home with no single element detracting from the others.

This panel siding with a skip-trowel texture characteristic of real stucco has grown very popular with half-timber styling, which also conceals panel joints. It comes either prepainted white, or primed for field painting. Photo courtesy of Masonite

A fresh Tudor-like half-timber look was given to the exterior of this home using small aggregate coated panels of plywood.

Tudor-style homes generally use several materials on the exterior. The home above utilizes Tudor styling in an all wood exterior. The home below shows a more traditional Tudor style.

phalt roof coverings, the 1½-inch nails may give adequate penetration into the sheathing. For old wood shingles, the nails for re-roofing should be 1¾ or 2 inches in length.

Have a sharp utility knife on hand, with extra blades, for cutting asphalt shingles. In conjunction with nailing, wear a carpenter's apron with pockets for holding a nail supply. To make shingle alignment guidelines, obtain a good-quality chalk-line and reel and two types of rules: a 6-foot retractable steel tape rule and a 6-foot carpenter's folding rule. You will need at least a gallon of roofing cement. This is a black asphaltic soft material that is spread with a broad 4-inch painter's scraping blade or a 6-inch drywall cement blade. Also obtain a small 4-to 8-inch painted trowel for applying cement.

With a roof whose slope is low enough to walk on without handholds, you may be able to do the re-roofing job with just one ladder for access to the roof. With steeper-sloped roofs, you will probably need a second ladder with a ridge-hooking device to lie flat on the roof slope. In both cases, the ladder should be rung-type. A single extension ladder may serve the purpose on 1- and 1½-story homes; on 2-story homes, get an extension ladder for access plus a single-rung ladder for roof use.

Don't use old ladders that have been lying around the garage or basement for years and years. One slightly cracked or rotted rung, and you're in the hospital with a broken bone, or two.

If you have much eave repair work, you'll save time and effort with a scaffold. Metal scaffold brackets that nail to the side of the house can be used with 2 x 10 or 2 x 12 walking planks. The brackets are reasonably inexpensive. An alternative may be found in some contractors' supply firms, which rent various scaffolding equipment.

For moderate to steeper sloped roofs, consider renting scaffold equipment not just for eave and lower re-roofing work, but also for delivering the roofing materials to the roof. On slopes of 5 or 6 inches rise per foot, shingle bundles can be lifted to the roof and the bundles spread out lying flat. They'll stay there. But on steeper slopes, the bundles tend to slide and get out of control. Flat scaffold platforms become a logical storage spot for shingle bundles on their way up.

When you purchase your shingles, cement and accessories, ask your supplier about foot scaffolds. A few retail building supply companies offer foot scaffolds for use by their shingle customers or do-it-yourselfers. If not available, a foot scaffold is easy to make and should be used on any roof whose slope is 6-in-12 or more. A foot scaffold is nothing more than a length of lumber with metal strips attached. Use a 2 x 6 that is 16 feet long. Nail 16- or 18-gauge

(fairly stiff) galvanized metal straps, about 2-to-3 inches in width, to the flat side of the 2 x 6. Three straps are used, one at midpoint and the other two at or near each end. The side of the 2 x 6 that the straps are nailed to is the down side; the strap length is about 14 inches, and the strap spacing should be such that they will fit between the normal nail spacing of asphalt strip shingles. The foot scaffold is put into position by slipping the straps under the top shingle course being worked on, bringing the edge of the 2 x 6 up uniformly to the shingle butts and nailing the straps in place with ¾-or 1-inch wide head nails. The foot scaffold gives the applicator a foothold, and it is moved up the roof and from place to place by lifting the shingles for nail removal and then renailing the scaffold at another location.

Safety provisions In all roofing work the danger of slips and falls is always present, even on low-sloped roofs. Roofing contractors and applicators become acclimated with their feet, and balance as sailors and boatmen do. They're used to working at heights and lifting bundles of shingles around. They're constantly aware of roof edges and roof obstructions in the form of plumbing vents and attic ventilators. The do-it-yourselfer is not acting instinctively as are roofers. The do-it-yourselfer must concentrate to keep within safety bounds and: (1) think safety in advance of every motion; (2) move more slowly than on the ground.

Good footing is very much a safety factor and will be helped by wearing soft-soled shoes. Rubber or crepe soles are a must. Avoid working when the roof surfaces are moist, such as during the morning dew period or on foggy, misty days. Stop work during short rain showers. Don't work when there are strong or gusty winds.

Use care in setting your access ladder. Select a central and convenient point for placing the ladder. See that the feet are firm and level. If on a sloping location provide blocking under the low side and stops to keep the ladder foot from slipping. Or use ladder-leveling devices on the feet. Another safety measure is to anchor the top in some way. One method is with the ropes that go through an open window and tie to a piece of 2 x 4 or 2 x 6 on the inside that stretches across the open window extending slightly beyond the two jambs.

In handling an aluminum ladder or in working around the side of the house where the home's electrical service is suspended, use particular care to avoid touching the electric wires. Serious shocks can result in an instant from the slightest touch.

One of the very real hazards in do-it-yourself roofing is the family itself. Keep your mate responsible for ground control. No shouting up to you to come to the phone. No "what do you want for lunch" com-

Footing holds for roof work are provided simply by temporarily tacking 2 x 4s in place. Photo above shows new plywood sheathing being applied over old roofing. At right, the roofing 2 x 4s are tacked through wood shingle courses, nail holes filled with caulking after removal. Some roofers use sheet metal straps slipped between shingles to hold the footing 2 x 4s.

munications. Kids and playmates must not only stay off the roof, but shouldn't be allowed around the ground area on the side of the house where you're working. Hand tools, shingles and other roof items can tumble down and cause serious injuries to those on the ground.

The learning roofer quickly picks up safety tips and applies them: Which foot and handhold to start down a ladder—Which position to place different tools on the roof so they won't slide—What not to trust, like roof gutters.

People vary in their feeling of safety while working. If you take it in stride, fine . . . just keep mentally alert. If, however, you're one whose thoughts at every moment focus more on safety than on the work you're doing, you may have to change your decision about doing the work yourself. A possible compromise for some might be the making of a safety harness. For the cost of 50 or 75 feet of good quality ½-inch nylon rope, you can fashion a harness to fit around the upper portion of your body, something like a life jacket. Such harnesses may be available ready-made at marine supply stores. If you knot loops at intervals in the rope before stretching it over the ridge, you'll have handy points at which the line from the harness can be secured to the loop with a safety type hook. Thus moving from one location to another, or getting up and down from the roof, simply involves unhooking from the loop while the anchor rope remains in place. Each end of the anchor rope is firmly fastened on both

sides of the ridge to an opened window block, a sill cock, a porch column or whatever may be convenient on the side of the house.

Stapling equipment Every do-it-yourself re-roofer ought to take time to make a telephone search for renting pneumatic stapling equipment if he's applying roof shingles on new construction, or to a roof deck from which the old roofing has been removed. The reason is speed. Applying roof shingles alone is a tedious chore that goes slowly. Even with two applicators, the task is still a bore. Light-duty pneumatic staplers can cut application time; just how much depends upon the nailing rate.

Some rate figures on expert roofers are available. Reports from actual roofing jobs have indicated that a shingler can do 25 percent more with a stapler than with a hammer and nails. Instead of 15-squares-a-day pace, these expert applicators can turn in 20 to 25 squares a day, and work can sometimes be done as effectively with one man as with the normal two-man crew.

Now, your time-saving with a pneumatic stapler could be even more because these workers are hammer-and-nail specialists. They work unbelievably fast to turn in 15 squares a day. Your pace with hammer and nails will be good if it is half that. But a stapler is a one-hand machine while nailing is a two-hand operation. Thus stapling is a method that will really facilitate and simplify your job.

Don't try the job with the manual-type stapler you might use for installing insulation. A pneumatic stapler that handles wide-crown, 16-gauge, 1-inch long galvanized or zinc-coated staples is needed. For many light-duty staplers, a small compact compressor of the type used for paint spraying is often quite satisfactory. Some staplers designed for roof shingling have integral shingle exposure gauges and carbide-tipped nose plates for long life in contacting the abrasive mineral-granuled surfaces.

While pneumatically driven staples are a shingle fastening developed in recent years, the method has gained acceptance. Underwriters Laboratories will now issue a Class C Label for staple-applied asphalt strip square-tab shingles on new roofs when 16-gauge staples are used.

The Asphalt Roofing Manufacturers Association has issued guidelines for staple application. Among them are:

- staple use limited to application of wind-resistant shingles with factory-applied adhesive;
- staples not recommended for re-roofing;
- staples to be 16-gauge, zinc-coated with minimum crown of ¾ inch and of length sufficient to penetrate ¾ inch into wood deck lumber or through a plywood deck;
- four staples per 3-foot shingle;
- use divergent chisel point staples.

One final item of equipment may or may not be needed, but homeowners will probably benefit from it: a rain-protective tarp or plastic film cover. Old-time canvas tarps run into substantial money for larger sizes. Heavy polyethylene film fabricated with grommets costs less. But large size sheets of plain plastic film are reasonably priced and can be tied down at corners and weighted on top with a few loose shingles. Such protection is most apt to be needed when the re-roofing work involves removal of the old shingles. Some roofing contractors prefer

Pneumatic-power staplers are increasingly being used on new home roof shingling and, to some extent, on re-roofing jobs. The reason is simple: time savings. The stapling method usually should be limited to application of seal-tab type, wind-resistant ashpalt shingles.

to remove all old shingles at once before proceeding with the new; the home's interior will be less subject to sudden rain damage, however, if an emergency covering is on hand and if the removal of old shingles is done a bit at a time—just a jump or two ahead of the new shingle application. (This latter method is really suitable only on lower-pitched roof slopes, where the clean-up following shingle removal will not disturb or damage the new shingles.)

Estimating and Ordering

The overall area of your roof in square feet is a summation of the areas of its various plane surfaces. Figure the area of each plane surface and then add them all together. To obtain these areas, you need edge measurements or close estimates of distances. It's fairly easy geometry for most folks. Keep in mind the ways of figuring right-triangular areas (height times width divided by two) and truncated rectangles having one side on an angle (width times length at midpoint of angled side).

Certain lengths in lineal feet will also be required including:

(1) eave and rake length for metal edging;
(2) valley lengths for roll roofing liner or sheet metal flashing;
(3) hip and ridge lengths for preformed hip and ridge shingles.

Roofing materials are ordered in units of 100 square feet or "squares." Add to the total area an extra 10 percent for cutting waste and starter courses. This extra amount should also allow you to come out with some left-over shingles that can be stored for possible future repairs.

With new roofing or on old roofs having old shingles removed, 15 pound asphalt impregnated roofing felt is required over the whole roof area. This is also ordered by squares and is applied with 2-inch headlaps and 4-inch endlaps. If old shingles are removed along eaves in order to make repairs on sheathing or fascia boards, their place can be taken by a layer of mineral surface roll roofing. This lower course of roll roofing, extended upward to a point about a foot inside the exterior wall, is recommended in new construction in cold-weather areas to help prevent moisture infiltration under shingles due to ice damming along the eaves. Sometimes mineral surface roll roofing is available in colors matching those of shingles and is used as a valley liner. And where not obtainable in matching colors, it can be turned mineral-side down for valley use.

Unloading materials Roofing suppliers are used to getting instructions from roofers on where the shingle bundles are to be unloaded. With low-pitched homes, the unloading is often directly to the roof surface. With your home, figure out where you'd like to have the bundles placed and then indicate these instructions with your order. Then, be sure you retain a copy of those instructions so that you can repeat them to the driver who delivers the shingles. Don't order them for delivery too far in advance of the time you're ready to work on the roof, and remember that suppliers normally expect payment within 30 days of delivery.

How to Estimate Quantities of Roofing

Computation of the roof area that needs covering is simple when you have a simple floor plan and uncomplicated roof. However, the method given here is suitable for all kinds of homes and roofs. It is the estimating procedure recommended in the application manual issued by the Asphalt Roofing Manufacturers Association. While the procedure may take longer than other estimating methods, it is accurate and allows an estimate from the ground without any roof climbing or roof measuring.

The method converts the house area at ground level to roof area, using conversion factors that depend upon the slope or pitch of the roof. First comes the determination of this slope or slopes. The span, rise, and run of a roof can be represented by triangles. In an equilateral (all sides equal) triangle, the base leg represents the span, the height of the triangle represents the rise of the roof, the equilateral legs represent the roof slopes or pitch, and half the base leg or span is what is referred to as "run."

For estimating purposes, the roof pitch or slope angle can be "measured" quite accurately using a carpenter's folding rule as shown in the drawing. One end of the rule is shaped into a triangle while standing away from the home. The rule is held at arm's length and the last two rule sections adjusted to parallel the roof slopes as indicated.

When the two sections have been adjusted so the view shows their angle is parallel to the two roof slopes, hold the overlapping end of the rule firmly on the base section. The reading point is shown by the arrow, sketch A. The reading is taken at the nearest eighth-inch point on the base section, as indicated in sketch B. From that eighth mark go down in the chart below to find the pitch fraction and the rise in inches per horizontal foot. Example: an end-of-the-rule reading of 23⅜ inches means the roof slope has a 4-in-12 pitch.

Ground area of the home is determined by on-ground measurements using a steel tape. The drawing on the facing page will be used as an example for the balance of the computations. This might be a typical dwelling with a few roof complications in the form of overhangs and dormers, plus ridges at varying heights. The projection below the perspective drawing shows the total ground area covered. In making the roof area computations, exterior wall measurements are taken at ground level but dormer and chimney measurements can be made in the attic space. Most dwellings can be measured up in this manner without need for climbing on the roof. After all measurements have been made and duly noted on paper along with a rough sketch and after the various roof pitches have been determined, you can proceed with the calculation of the different roof portions.

In the example home, the likely starting point is the main roof, whose slope is 9-in-12. This will be followed by the 6-in-12 portion. Then, allowances are made for the duplicated roof areas and for such additions or subtractions as those for chimneys and dormers.

Slope conversion is necessary because the roof area over a ground area is larger than the ground area. A one-plus conversion factor for various roof pitches can be calculated, so that a simple multiplication changes the ground area into the roof area desired.

Later, after roof area has been determined, it will also be necessary to estimate the length of hips, ridges, and valleys. Ridges, being parallel to the ground, are no problem. But hips and valleys also slope and need conversion.

The conversion figures below provide factors for more common roof slopes. The horizontal ground area or distance is measured or estimated in square feet or length in feet; multiply that by the conversion factor for the appropriate roof pitch, and you come up with the roof area or hip/valley length.

Adjustments to roof area calculations are needed for supplementary areas such as the single dormer in the example house or places where one roof area may project over another because of eave overhangs. There are also adjustments to be made where the overhangs, eave or rake, extend beyond exterior walls. Even a short overhang can mean a substantial area to be added with some homes.

In the example drawing, just a 4-inch eave overhang with duplications accounted for will add about 12 square feet of area for the duplicated portions, before conversion to roof area. After all roof areas have been calculated and the summation made to reach a total area, add about 10 percent more area to cover wasted material. Roofing shingles are sold by the square, that is per hundred square feet. Divide your total area by 100 to obtain the number of squares of material required.

Accessories need to be estimated in much the same way, beginning with the hip-valley-ridge lengths needed for hip-ridge shingles or flashings for val-

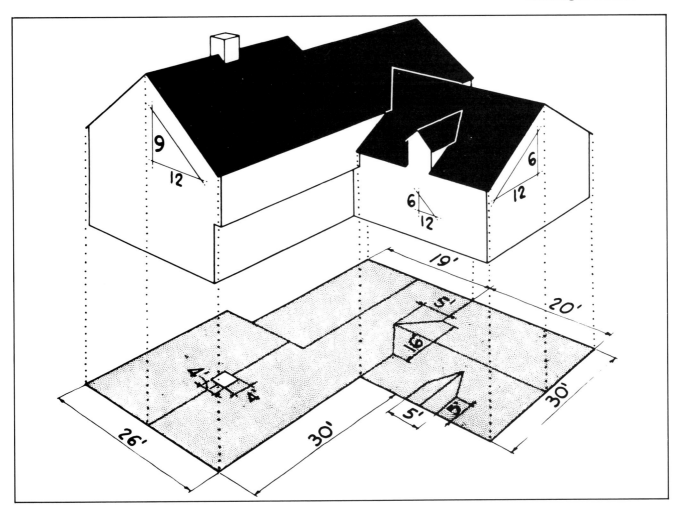

A

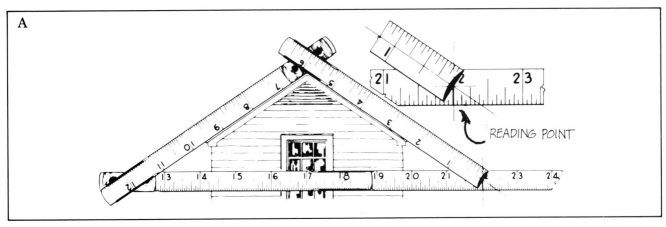

B

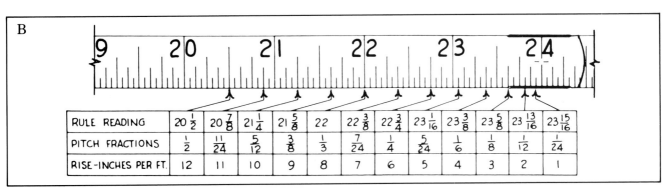

RULE READING	$20\frac{1}{2}$	$20\frac{7}{8}$	$21\frac{1}{4}$	$21\frac{5}{8}$	22	$22\frac{3}{8}$	$22\frac{3}{4}$	$23\frac{1}{16}$	$23\frac{3}{8}$	$23\frac{5}{8}$	$23\frac{13}{16}$	$23\frac{15}{16}$
PITCH FRACTIONS	$\frac{1}{2}$	$\frac{11}{24}$	$\frac{5}{12}$	$\frac{3}{8}$	$\frac{1}{3}$	$\frac{7}{24}$	$\frac{1}{4}$	$\frac{5}{24}$	$\frac{1}{6}$	$\frac{1}{8}$	$\frac{1}{12}$	$\frac{1}{24}$
RISE-INCHES PER FT.	12	11	10	9	8	7	6	5	4	3	2	1

leys. Eave and rake lengths will be added to give the total lengths needed of metal edging. The amounts of nails and roofing cement required will depend upon the nature of the job, rather than the areas. Obtain guidance from your roofing supplier at the time that materials are ordered.

Conversion Table		
Pitch Rise (inches per foot)	Roof Area Factor	Hip/Valley Factor
4	1.054	1.452
5	1.083	1.474
6	1.118	1.500
7	1.157	1.524
8	1.202	1.564
9	1.250	1.600
10	1.302	1.642
11	1.356	1.684
12	1.414	1.732

Asphalt Shingle Application

Eave and rake roof sheathing edges receive initial attention when doing preparatory work for shingles. Inspection for decayed sheathing boards and the removal of shingles along eave edges has been mentioned, the point being that best roofing results follow when metal drip-edge strips are installed along all eave and rake roof edges. These metal strips need solid and firm wood underneath to which they can be nailed. If sheathing along the edges is not in first class shape and dry, it should be replaced along the very edge by a 1 x 4 or 1 x 6.

The lack of metal drip-edging under the shingles permits water draining down the shingles to seep back under the eave and rake edges, and keep the wood sheathing below almost continuously wet or moist. And dampness facilitates rotting and decay. The absence of metal edging is probably one of the most important causes of damaged sheathing along these roof edges.

If your roofing work is on a new home where there are no old shingles to be dealt with, or on a roof where all old shingles have been removed, then your metal drip-edging installation should be coordinated with the laying of felt underlayment. The metal edge along eaves is applied first and overlaid by felt, but the edge on roof rakes is installed over the felt. Normally, roofing contractors do not apply the felt underlayment over the entire roof before starting the shingling. They prefer to unroll several rolls of felt beginning at one roof rake, the rolls being properly headlapped. Then shingling is begun, and the shinglers work their way upwards and to the right. As shingling progresses, the felt rolls are unrolled a little further to the right and another

1. First re-roofing strip along eave will be a 5 x 36 inch strip, so this width is measured and cut so that the 5-inch strip has the black adhesive seal-down stripping near the bottom.

roll or two added at the upper left. This amounts to a diagonal procedure in application and seems easier for handling and placement of shingles. Felt underlayment need not be regularly fastened, requiring only scattered nails or manually applied staples to keep the wind from lifting it.

The normal procedure on hip roofs is different. Here, application usually starts at the center of a roof plane and shingle laying is done in both directions. Full details on use of starter strips, shingle positioning, points of nailing, and use of alignment chalk lines is provided in the illustrations. However, the reader should be warned that most of the instruction material deals with the common three-tab shingle. With many newer types of shingles on the market, the application details vary somewhat with the specific brand of shingles being used. It is important therefore to read carefully the instruction sheets that come with the shingle bundles. This is particularly true in order to understand how shingles should be properly staggered, course to course, and how such features as alignment notches or slits are utilized to provide accurate line-up of strips in each course.

Low-slope roofs A very large number of homes have relatively low-pitched roof slopes ranging from 2-in-12 up to 4-in-12. This lowness of pitch contributes to roofing problems in northern sections of the country where average January temperatures run 25°F or less. These roofs require some special treatment to insure roof durability. The special treatment suggested by most roofing manufacturers is a double felt application, cemented down with roofing cement.

The 15 pound roofing felt is 36 inches wide, but a strip 19 inches wide is cut as a starter strip along the eave edge. Then, a full-width 36-inch strip is cemented down over the starter strip. Then additional

2. *Balance of the cut strip can then be further trimmed down 2 inches to provide another 5-inch wide strip with seal-down adhesive, which can be applied later at the ridges.*

3. *Beginning or starter course of the 5-inch wide strips is then positioned against butt of the second course of old asphalt shingles, with seal-down stripping near the eave edge.*

4. *Nailing of the starter strip takes four nails per strip, placing the nails at a point just below the seal-down adhesive.*

5. *Second course of shingle strips is then placed, the strips having been trimmed down to 10-inch widths. This course is positioned butting to the third course of old shingles.*

strips are lapped 19 inches over the previous strip and cemented down. This undershingle treatment is sometimes called "eave flashing" when applied on slopes of 4-in-12 or 5-in-12 roofs. For this use, the double felt layer is extended up the roof to a point that is 2 feet inside of the exterior wall (projected vertically upward), or the distance of the roof overhang plus 2 feet. The cementing-down process is done with a rectangular cement finisher's trowel using cement at the rate of about two gallons per 100 square feet.

The same double-felting and cementing down is suggested by some roofing manufacturers for best results in applying shingles to lower slopes of 2-in-12 to 4-in-12. For these applications, however, the felt layering and cementing is carried out fully up the slope to the ridge.

Nailing shingles The general rule for nailing is four nails per strip shingle, each nail spaced about 11 or 12 inches apart depending on style or type of

6. *Third and following courses are then placed conventionally, each butted to an old shingle course. This series from a complete installation set covering Johns-Manville fiberglass shingles.*

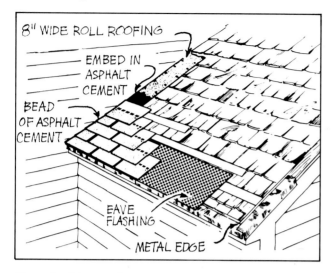

Eave flashing and sidewall flashing with roll roofing cemented down is desirable when new asphalt shingles are being applied over old roofing.

Edge shingles that meet a sidewall or the rake of the roof can be precut quickly by using a template shingle that's a half tab in width.

Diagonal application makes for easier shingle handling according to expert roofers; this method is preferred, rather than applying full shingle courses across the roof horizontally.

shingle. Nail locations on a three-tab shingle are just above the cutouts and just below the line of adhesive stripping. Check nailing recommendation by individual shingle manufacturers.

In nailing down the shingles, don't let the nail-head indent the shingle. Drive the nail until the head is just barely flush with the shingle surface.

As additional shingles are applied, keep constantly alert for alignment of the courses, keeping the butt lines straight and parallel and also parallel to roof eaves. Many producers of felt underlayment paper provide white lining on the black paper that is intended to help with this alignment. When re-roofing over old shingles, alignment with old butts may help in aligning the new shingles. Where there are no such guides to strip alignment, it's advisable to snap chalk lines every other course, 10 inches

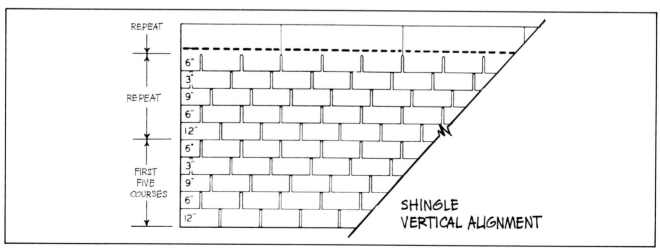

Random spacing of strip shingles is achieved by removing different amounts from the rake tab of succeeding courses. Sketch shows amounts of cuts, minimum being 3 inches. With a five-course repeat pattern, the eye tends

not to follow the vertical alignment of cutouts and the spacing appears random. Cut-off pieces may be used at opposite end rakes or at sidewalls.

Cementing down of new shingles all around the chimney should be done whether old counter-flashing is used or not. If not, then additional cement caulking should be applied along the joint between shingles and the chimney faces.

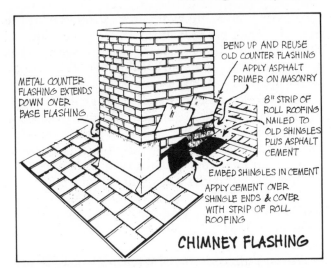

CHIMNEY FLASHING

Chimney flashing when re-roofing can often make use of old counter-flashing when the old material is still serviceable.

apart for 5-inch shingle exposure. Similar chalk lines, snapped vertically, can keep shingle tabs aligned vertically if this is desirable. On some roofs, a random pattern of tab positions is followed.

Nailing practices vary somewhat with very steep slopes, such as those with mansard or gambrel roofs. If the slope is steeper than 21-in-12, check your shingle manufacturer's steep-slope instructions for nailing and for application of cement spots under the tabs.

One word more about nailing. Shinglers do not hold the individual nails between forefinger and thumb with palm down. They hold the roofing nail between the index finger and the long middle finger with the palm up and back of hand resting on the shingle. Why? Nobody seems to know for sure, and you'll get a different answer from nearly every roofer you ask. But that's the way they do it, and you might want to do the same.

Roof valleys These are main drainage channels and need a bit of special attention. Shingles come together here in their coursing at right angles, and the strips have to be cut on an angle before being nailed down. A valley flashing material is applied to the deck before shingling on new construction or on re-roofing jobs.

To describe each method briefly: an open valley is one in which the angled shingle edges are kept 6 to 8 inches apart, leaving the top surface of the flashing exposed as a main drainage channel; a closed valley has no such opening and the flashing material is completely concealed.

On architecturally designed and other higher-priced homes, metal flashings are common though their use is limited usually to new homes (and they are not so commonly used as they once were). Reason: use of roll roofing is just as effective and often easier. Ordinarily, 18 to 24 gauge galvanized sheets cut to 24 inches in width are nailed down along the valley, keeping nails at the edge and slightly pre-

Sidewall juncture, the point where a roof surface meets the siding of an upper story, also requires the use of a roofing-felt underlayment strip of flashing plus the cementing down of the adjacent shingles, as indicated in this photo.

bending the metal to fit the valley angle. Upper strips overlap lower strips by about 10 inches, and good practice calls for cementing down of the lapped portion.

Roll roofing has become the commonly used flashing material for both open and closed valleys. For open valley installations, two layers of roll roofing are applied, the bottom layer being a strip 18 inches wide with the mineral surface down. The edges of this under strip are coated with roofing cement and then the top strip, 36 inches wide, is run down the valley over the lower layer. If a lap is needed, the upper strip overlaps the lower strip by 12 inches and is cemented down. The flashing is then marked with a chalk line marked equally on each side of the valley. The usual practice is to begin at the top or ridge with 6 inches between marks, and allow for gradually increasing width down the valley at the rate of ⅛ inch per foot. If it's 16 feet from ridge to eave, the chalk line marks will be 6 inches apart at the top and 8 inches apart at the eave. The shingle edges are then cut to these marks.

Closed valleys can be handled either of two ways.

Closed Cut Method. Shingles are laid up the slope on one side of the valley with each course crossing the valley and extending on the other side by about a foot. Nailing is kept 6 inches away from the valley center. Then shingles are applied on the other side of the valley. At the valley center, these shingles are cut to the valley angle and their valley-angled end then embedded in a 3-inch wide bed of roofing cement.

Woven Valley Method. Increasingly used in preference to the above technique, because it doesn't require shingle cutting and cementing, shingling is done on both sides of the valley, and the courses are alternately carried across the valley. The extension up the opposite slope is at least a foot, and nails are kept back 6 inches either side of the valley center.

On re-roofing work where old shingles have been removed, the foregoing valley flashing or shingle crossing is done over the building felts. Where old shingles remain in place on the roof, a little extra work is necessary in the old shingle layer for open valleys. These open valleys need to be filled level with the top surfaces of the old shingles. If the old shingles are asphalt, the filling can be a double thickness of mineral surface roll roofing cut into strips to fit the valley and cemented down in random spots about 8 or 10 inches apart. If they are wood shingles, the open valley is a bit deeper and

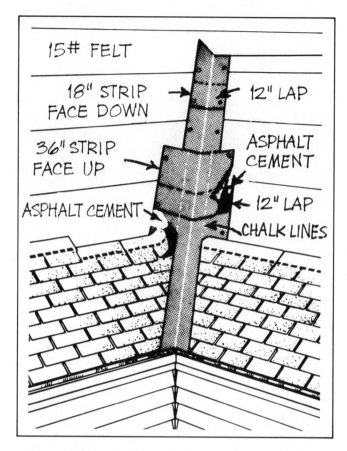

Open valley method uses two layers of mineral-surfaced roll roofing and is suitable on jobs where old roofing material has been removed.

the wood strips, one each side of valley center, is indicated. Again, cementing down is suggested, although occasional nails will probably be needed to hold the strips in place. Keep the nails to the edges away from the valley center. After filling the old valley, proceed with new valley flashing work.

Hips and ridges These crowned junctures of roof planes are both handled in pretty much the same way. Regular roof shingle courses are brought up to the ridge or hip line and cut along that line. Then a series of lapped single tabs are applied. Along ridges, the lapping can be in either direction. Along hips, the lapping of course is in weather fashion, upper units overlapping lower ones.

Hip and ridge work has been simplified in recent years, with many shingle manufacturers furnishing preformed hip and ridge shingles. This eliminates the need to cut three hip-ridge units from a three-tab strip as well as the troublesome effort to conceal corners of the cutouts with the cut tabs. If cut tabs are used, each unit should be bent in the center before applying to the hip or ridge. Nails one inch from the shingle tab top on each side of the ridge or hip are driven, exposure is usually 5 inches, the same as that on roof plane shingles.

Dormers What workmen call "fiddling around"

Ridge shingles are single-tab widths cut from shingle strips and applied with same exposure as used on field shingles. Use one nail each side of the ridge. With hands, carefully prebend each ridge shingle. In some areas, suppliers offer precut ridge and hip shingles. It is good practice to cut ridge shingles with a slight taper on the back side.

is necessary where roof dormers occur. Roughly translated, this term means some extra time is needed to handle the proper fitting of materials.

The problem area is the dormer valleys, and particularly the lower end of the valleys where they join the main roof slope. The recommended way to handle shingles at these points is, briefly:

(1) Keep laying the main roof shingles up the slope on either side of the dormer sidewalls until the juncture of the dormer roof with the main roof is reached. The last course top edges should be just above the juncture and the strip ends each side of the dormer must be cut so the strip fits under the dormer eave trim and bends slightly upward on the sidewall.

(2) Then the valley flashing for the dormer valley is applied and trimmed at the lower end to overlay the main roof last shingle course at a point near the top of the cutouts.

(3) On the dormer side, the flashing is trimmed to just beyond the eave edge.

(4) From this point on, the procedure for dormer valley shingling is like that for an ordinary open valley.

At the dormer peak, a little more special attention is needed. Before reaching peak level, a chalk line should be snapped measuring up from the last main roof shingle course the proper number of intervals.

The object is to have the chalk line in position just above the dormer peak so that shingling on either side of the dormer will be brought up evenly for later continuation of the courses above the peak.

Miscellaneous shingling tips

- If closed cut or woven valleys are used on main roof junctures, they can also be used on dormer valleys to maintain a uniform appearance. Just use care in fitting at the lower end of the valley and under the dormer eave.
- In bending hip or ridge shingles, be careful. Use the edge of a 2 x 4 or the roof ridge itself to start the bending and remember that a little warming up of the shingle tabs will make them more flexible, easier to bend, and less likely to crack or split.
- A trick of some roofers may be worth copying on heavily textured roofs: double up the hip and ridge shingles to give a thicker look, more in keeping with the texture. Simply use two hip or ridge shingles in place of one and nail them in place together.
- Slight warming and prebending will help to keep the cross-valley portion of shingles lying flat when using the woven valley method.
- On steeper roofs the water run-off is much faster, and this cuts the amount of water a valley channel has to handle. As a result, open valleys are not often used on the steeper pitches of roofs such as the 10-in-12 or 12-in-12 slopes common on

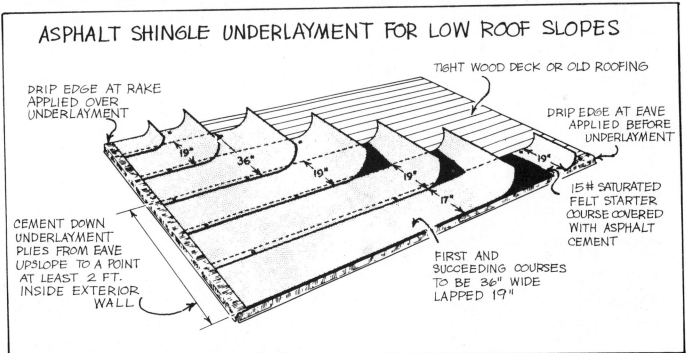

ASPHALT SHINGLE UNDERLAYMENT FOR LOW ROOF SLOPES

DRIP EDGE AT RAKE APPLIED OVER UNDERLAYMENT

TIGHT WOOD DECK OR OLD ROOFING

DRIP EDGE AT EAVE APPLIED BEFORE UNDERLAYMENT

CEMENT DOWN UNDERLAYMENT PLIES FROM EAVE UPSLOPE TO A POINT AT LEAST 2 FT. INSIDE EXTERIOR WALL

15# SATURATED FELT STARTER COURSE COVERED WITH ASPHALT CEMENT

FIRST AND SUCCEEDING COURSES TO BE 36" WIDE LAPPED 19"

Low-slope application of asphalt shingles ranging from 2-in-12 up to (but not including) 4-in-12 should have a cemented down felt underlayment as indicated in this drawing. According to the Asphalt Roofing Manufacturers Association, the cementing-down should extend up the slope to a point at least two feet from the inside face of the exterior wall. Asphalt roofing cement is applied at the approximate rate of two gallons per hundred square feet between deck and felt, and between felt layer and felt layer. Use a comb-type trowel to spread the cement.

expansion attic or Cape Cod type homes.

• When applying the newer irregular-butt or double-layered shingles, be sure to check the instruction relating to alignment devices incorporated into the shingle. These shingles usually have slits or notches of some kind to help keep adjacent strips in alignment.

Flashings This subject is an important adjunct to roofing on new construction. It is one of secondary importance on re-roofing jobs (except for the eave and valley flashings already mentioned). This is because the flashings on older homes ordinarily remain in good condition and do not need replacement. However, there are some homes where new or supplementary flashing work is advisable; details are provided in later chapters.

One word of advice at this point to the do-it-yourselfer, however. Use care in working on the roof not to disturb present flashing materials around chimneys, vent pipes, and ventilators. A slight displacement of a piece of flashing material can cause a leak in the new roof you have added.

Wood Shingle Application

What has been said previously with respect to asphalt shingle roof preparation applies, with no practical differences, to preparation for wood shingles, new construction, or re-roofing. You need a firm and even base. You need rake and eave sheathing or roof edges in good firm condition. You need metal drip-edging along these edges. You need to fill old valleys. You need to remove old hip and ridge shingles and perhaps use thin wood strips or lengths of thin bevel siding at hips and ridges to bring the base up to overall evenness.

Wood shingles on new construction can be applied over spaced wood sheathing boards. This type of sheathing involves the use of square-edged 1 x 6s which are spaced on centers equal to the weather exposure at which the shingles are to be laid, but not exceeding about eight inches. Spaced sheathing is usually preceded by three, four, or five courses of solid boards along the eaves, sufficient to come up to a point about a foot inside the exterior wall. Handsplit shakes can also be applied with spaced sheathing boards.

For re-roofing applications, wood shingles can be applied over nearly any kind of old roofing material except the harder ones, such as slate, asbestos shingles, or metal roofing. The principal difference between applying the shingles for re-roofing or new construction is the use of longer nails: 3d (penny) and 4d nails 1¼ and 1½ inches long are suited for new construction shingling; re-roofing applications use 5d and 6d, 1¾- and 2-inch long nails. The 4d and 6d nails are for 24-inch long shingles, the others for 16- and 18-inch shingles. All nails should be rust-resistant.

Tools needed About the same equipment in ladders, roof jacks, scaffolds, and hand tools mentioned for asphalt shingles, also applies for wood shingles. But the cutting is different. Most roof shinglers use a roofer's hatchet for nailing instead of a hammer because the hatchet's cutting edge is well suited to the most common form of cutting—splitting the shingle to the width needed.

For butt alignment, the usual device used with wood shingles is simply a straight square-edged board, 1 x 6 or 1 x 8, 12 to 16 feet in length. The straight-edged board is tacked in place above the previous course at the correct exposure marks and the next course of shingle butts placed against the straight edge. Where irregular butting is done, the straight-edge can still be used as a horizontal and parallel guide for the lowest shingle butts.

Estimating and ordering of wood shingles is much like that described for asphalt shingles. Obtain your overall roof area and 10 percent waste allowance. Now, the normal range of shingle exposures recommended for different pitched roofs varies slightly (see below), but in each case, the application provides three layers of shingle thickness at any point on the roof. But regardless of shingle length and exposure, all wood shingles come bundled so that four bundles will cover a square of roof area. Allow an extra square for each hundred lineal feet of hips and valleys.

There are certain limitations which should be mentioned in the use of wood shingles. They should not be used on homes whose roof pitch or slope is less than 3-in-12. No. 1 grade shingles are the ones whose application is described in these pages. With lesser grades, reduced exposure is suggested, and inquiry should be made to the supplying mill or the Red Cedar Shingle & Handsplit Shake Bureau for application instructions.

The suggested shingle exposures for different shingle lengths are as below:

	Shingle Length		
Roof Slope	16″	18″	24″
3-in-12 up to 4-in-12	3¾″ exp.	4¼″ exp.	5¾″ exp.
4-in-12 or steeper	5″	5½″	7½″

Hips, ridges The installer must become adept at beveling the side edges of shingles. Uniform width shingles are used and while there are other methods of joining at the ridge or hip line, the preferred method is called the "Boston Hip." It involves a different facing overlap on successive courses. It's sometimes called "lacing" or "staggering." A pair of shingles match each other on either side of the

1. Shake application begins with 15 lb. roofing felt layers which will interleaf with shakes. At left, starter shake course is begun while (right) the first course is laid directly over starter units.

2. Shake nails should be long enough to penetrate well into roof sheathing. Top end is tucked under felt strip. Space shakes about ½ inch apart, and keep joints in adjacent courses above and below offset by at least 1½ inches.

3. Valley flashing is of at least 26 gauge galvanized sheet metal painted in roof color with anti-rust paint. The 1½-foot wide flashing metal has a center crimp to facilitate drainage flow, and is nailed in place along outside edges.

4. Cut shakes along valley; a chalk line snapped on the metal will provide guidance for accurate shake cutting. Open valley between shakes should be a minimum of 6 inches wide at the top with a slight widening to 7 or 7½ inches at the bottom.

5. Hip and ridge shakes are cut right at the hip or ridge line (far left photo) and joint is then covered. Using factory-assembled hip or ridge shakes (near left) saves much cutting and trimming time.

Starter-finish shakes are available that can save on shakes and cutting. Photos courtesy of Red Cedar Shingle & Handsplit Shake Bureau

ridge or hip. One pair will have the righthand shingle overlapping the side edge of the lefthand shingle. On the course above and below this pair, the pairs are positioned so the lefthand shingles overlap the righthand ones. In each set of pairs, the side edges are trimmed to fit flush with the other shingle.

You can solve this tedious trimming work by spending a bit more money for the purchase of preformed, factory-assembled hip and ridge units, but these convenient shingles may not be available from all shingle suppliers.

Valleys The common practice on wood shingled roofs is for open valleys on low to moderate slopes, and closed valleys on steeper slopes such as the lower ones on mansard and gambrel roofs. With wood shingles or shakes, galvanized metal flashing is commonly used in valleys, 18 to 24 gauge metal in 20-inch widths, preferably with a center crimp. Many wood shingle installers prepaint the valley metal. It should first be cleaned with a solvent to remove the oily film on the metal. Then, use a heavy-bodied lead-oil or bituminous paint, allowing metal to dry before applying paint to the roof. On re-roofing, old valley channels may need filling as described for asphalt shingle valleys.

The thicker and coarser wood shakes are applied almost the same way as regular wood shingles. The principal difference is that roofing felt, instead of being applied as an underlayment, is applied interleaving with the shake courses. This interleaving is recommended when the application is on new or old wood sheathing, and also when shakes are applied over old roofing materials.

Shakes are available in 18- and 24-inch lengths, and the suggested exposures are 7½ and 10 inches, respectively. As with wood shingles, a doubled-felt course is desirable along the eaves and the first shingle course should be a doubled one. After each course of shakes is applied, an 18-inch (half the width of a roll) strip of 15 pound asphalt felt is applied over the top portion of the shakes, extending onto the sheathing or old roof surface. The bottom edge of the felt should be located at a point above the shingle butts of the course that it covers, about twice the exposure length. Example: with 24-inch shakes applied with 10-inch exposures, the felt strip bottom should be about 20 inches above the shingle butts over which the strip is applied.

Shakes are spaced further than shingles. Increase the side gaps to about ½ inch. As with shingles, the joints or gaps in alternate courses should not be in direct vertical alignment. With shakes, the common laying practice differs for some roofs from that used for wood shingles. Where straight-split shakes which have about equal thickness throughout are used, the "froe" end of the shakes is smoother than the other end and should be laid toward the ridge. "Froe" refers to the splitting device used, and this is the end of the shake from which the splitting has been done. Such positioning will insure a tighter roof.

Wood shingles and shakes can be laid in other ways and still provide just as serviceable and resistant a roof covering. Different appearances can be obtained by using, for example, 24-inch shingles on lower portions of a roof and 18 inches on the upper portions, a method that tends to lengthen the apparent distance from eaves to ridge. Shingles and shakes may also be applied with staggered butts. Roof shingling contractors can probably advise you of other ways that shingles and shakes can be handled. For the do-it-yourselfer who desires to be more individualized in his shingle or shake treatment, the counsel here is: go slow, think your design out carefully and make notations and sketches. Then inquire of the previously mentioned Red Cedar Shingle & Handsplit Shake Bureau what they think of your idea.

Supplemental Roof Work

This chapter is a round-up, in differing degrees, of roof-related jobs, some or all of which may need to be done on some older homes.

Some of the following items may be "musts" involving almost complete replacement, while in other cases a rather simple reconditioning will bring the old materials up to par. A few options the homeowner may desire because of the convenience in doing them just before the roofing is applied or repaired.

What's said in this chapter about chimney and other flashings, and work involving the flashings, should be accomplished before the roofing materials are applied. Also, if any tuckpointing of masonry chimneys is needed, it should be completed before the roofing in order to avoid later extra walking or traffic on the new roofing material.

So, bear with these varied topics. Take each item and consider whether or not it is applicable to your home. If so, place it in the proper mental niche where it belongs . . . fore or aft of its present position here.

Check Roof Flashings

Flashings have been previously covered for roof valleys. As was evident from that explanation, a flashing is an extra amount of material of one kind or another, positioned at certain critical points where leakage often occurs. Flashings are intended to prevent such leaks.

One critical flashing is where masonry chimneys penetrate the roof deck. Other common roof-penetrating elements are plumbing vents and roof ventilators. In recent years, residential skylights have grown popular; skylights mount on wood-framed roof projections and the projections are flashed. There may also be roof blocks as bases for TV antennas, and roofing shingles must be worked around these blocks with flashings.

Two-part chimney flashings Chimney flashings are always of metal, usually of copper because of its easy workability but also of galvanized sheet metal on lower cost homes. When re-roofing old homes, chimney flashings usually need not be replaced because the metal does not deteriorate.

Unless the home has had roof leaks in which chimney proximity might indicate flashings as the possible leak source, simple realignment of the top or counter-flashing members is probably suffi-

cient. Sometimes these top flashing pieces work loose because of shrinkage in the mortar joints, which brings mortar fall-out and the need for tuckpointing.

The lower portion of the chimney flashings—those pieces which fit the right angle or sloping angle between the roof deck and the chimney face—rarely work loose and seldom need attention.

What usually happens in re-roofing contractor practice is that the ends of the new strip shingles are cemented down adjacent to the old flashing material. Where the lower flashing is carried up the faces of the chimney only a short distance, some contractors will bend the top or counter-flashing material upwards and apply cement under the shingle ends and up the wall face, so that the cement will be mostly covered when the counter-flashing is bent back down into its former position.

On new construction, chimney flashings are one of those items that may be installed by different trades, depending on the locality of the home. A chimney mason, roofer, or sheet metal man may do the installation. No matter who furnishes the metal flashing material, the shingler is the logical man to

Aluminum foil-faced sealing tape can be a real time-saver in re-roofing for flashing around plumbing and chimney roof projections. Above, the foil-faced sealer, which comes in rolls of varying widths, is pressed into place around the base of a plumbing pipe.

do the lower or base flashing, and the mason or shingler to handle the counter-flashing, which has its top bent edge inserted ¾ inch into mortar joints in the masonry. The objective is to allow water to run off the chimney without getting behind the flashing material.

Sidewall flashings Where the roof of a one-story portion of the home meets the sidewall of a two-story portion of the same home, sidewall flashing is needed at the roof-wall juncture. This flashing is also of sheet metal cut into rectangular pieces, like those for the sides of chimneys, and also interleaved with the shingles in the same way. However, on new construction, the vertical face portion of the flashing pieces should be along the surface of the wall sheathing and later covered by the wall siding.

On re-roofing work, these sidewall flashing pieces normally remain in good condition. The common practice is not to touch them, but to apply new shingles with roof cement under their ends adjacent to the sidewall. Care should be used to avoid getting this hard-to-remove black cement on the face of the siding. If removing old shingles, be careful not to disturb or damage the flashing pieces, allowing them to be fitted in with the new shingle coursing.

Pipe flashings Soil stacks and vent stacks penetrate the roof and with each penetration, a metal flashing is used. On new homes, these pipe flashings are installed by plumbers when the final pipe lengths through the roof have been installed. They are preformed one-piece metal units that slip over the pipe end and have a square flat portion that lies on the roof slope. Under normal roof conditions, these pipe flashings almost never wear out or deteriorate, but if someone is careless when walking around a roof, they can be kicked and damaged, perhaps sufficiently to develop a leak. So, cue yourself and any roof contractor not to disturb or damage pipe flashings.

Approximately the same applies to roof ventilators. Hooded metal vent units may fit into either round or rectangular openings in the roof sheathing, and the vent units come with flat marginal metal extensions on all sides that serve as roof flashing. For re-roofing on homes having such vent units, exercise caution so as not to disfigure the flashing extensions.

Where old roofing materials remain in place the new roofing shingles are applied right up to the pipe or ventilator units and cut out to fit closely around them. Good practice calls for application of roofing cement on the old roofing, in a margin of about 3 inches, before placing the new shingles down to fit around the penetration.

A further consideration on some roofs relates to roof-mounted TV antennas. Some homes have their antenna towers mounted directly on the roof, usually with tower legs resting on angled blocking to fit the roof slope and to provide a level base for the legs. Often, this blocking is simply placed in mastic over the old roofing material and spiked to the roof framing. When re-roofing, this base blocking need not be removed or altered, but the new shingles should be turned up at the blocking edges and cemented to them.

Tuckpointing Chimneys

Sometimes homeowners believe they have a roof leak that is in actuality a chimney leak. Masonry chimneys, sturdy and permanent though they appear, do deteriorate with cracking, mortar shrinkage and fall-out, and development of openings in the chimney cap through which water can seep to the chimney's interior and down into the home.

The corrective measure for most such chimney difficulties is called "tuckpointing." It consists of chipping out loose mortar from the joints and troweling in fresh mortar, which is then tooled to a smooth finish. This work is relatively easy and well within the capability of the do-it-yourself homeowner. A small masonry chisel that will fit between the brick courses is used to chip out loose mortar to a depth of about ¾ to 1 inch. A premixed mortar cement can be obtained at building supply yards; the mix simply requires adding water and a bit of mixing with a hoe or shovel; then it is ready to use. Using a small pointed trowel or a long narrow tuckpoint trowel, the joints are refilled and after the mortar has hardened slightly, joints are tooled with a joint-finishing tool.

If a chimney has loose mortar joints which need tuckpointing, it will also probably have loose counter-flashing members, the top pieces of metal flashing that insert into mortar joints. Any loose joint fill material here should also be chipped or raked out, but don't use mortar to refill the joint. Instead, use a caulking material applied with a cartridge gun. The same caulking material should also be applied to cracks or openings in the chimney cap. Be sure to check the underside of the cap's projection beyond the chimney face. This is a place where cracks or mortar fall-out frequently occurs.

Ventilators and Skylights

We have already mentioned handling new shingles to fit around the bases of ventilators already in place. Where there are such roof ventilators for attic spaces, the home was probably built after proper attic ventilation requirements were written into building codes and standards. However, many older homes were built before such provisions; their attic spaces may have no ventilators at all in the roof, in the gable-ends, or under eaves. Or, if there are just a few vent openings, the area of the vent space may be insufficient for the size of the home.

Attic ventilation is not a subject to be brushed aside. It is particularly important in older homes, where the lack of such ventilation has caused decay damage due to moisture condensation in the attic space. If there is any indication of moisture, such as discoloration or sogginess in the roof sheathing boards or ceiling joists and insulation, lack of ventilation is a probable cause.

Ventilation of attic spaces is brought up here, however, because the homeowner may desire to update his home by providing better ventilation. And when re-roofing work is about to be done, it is a convenient time to install roof vent units—either the simple metal hood type, or electric-powered ventilators that furnish a positive measure of cooling to ease air-conditioning loads.

This is also a convenient time for a homeowner to consider another type of home improvement—a skylight to provide daylighting and possibly ventilation in the attic or other rooms. There are prefabricated aluminum and plastic-glass skylights available in a wide range of sizes. Some use flat acrylic plastic glass; others use domed lights of the same material. Some are fixed, while others can be opened like a window to allow fresh air in. They are frequently and primarily used in home remodeling jobs for adding daylight to attic, bathrooms, or stairhalls.

Skylights are normally installed on a wood-frame box that projects a uniform 4 to 6 inches above the roof surface. This is commonly called a "curb," and the box can be framed to skylight size using 2 x 4s or 2 x 6s. The roof opening for the skylight is cut in the roof sheathing with one side adjacent to a rafter. After cutting, rafter-sized members are cut to frame out the opening on the underside of the roof sheathing so the opening's edge is completely supported below by the framed members. Then, the skylight box or curb is placed atop the roof sheathing, old roofing having been cut back to allow placement of the box directly on the sheathing along the edge of the opening. The box or curb is then toe-nailed to the framing below. Metal flashing strips are then placed on the four sides of the curb beginning with the low-slope side, the sides, and then the top side. When the shingling reaches the curb, shingle ends should be cemented down adjacent to the curb. This skylight installation is quite easy for do-it-yourselfers and can be done on many homes from inside the attic.

Other Work

The most common job supplemental to re-roofing, and done after the new roofing is in place, is the installation of new roof gutters. On most existing homes, the guttering falls apart before anything else on the home's exterior. Most roofing contractors also do reguttering work, and it may be the first item mentioned when you call in a contractor to look at your roof. The installation of a new roof drainage system is increasingly becoming a do-it-yourself home-improvement job. Aluminum and vinyl plastic material manufacturers have developed roof drainage systems and accessories that simplify the installation of new gutters and downspouts. These are the same manufacturers who produce aluminum and vinyl plastic sidings, soffit panels, and fascia cover panels.

Reguttering work by roofing contractors is more apt to use conventional galvanized or aluminum material that must be soldered or brazed where sections join or connections are made. This soldering or brazing work is done right on the job, usually on the ground before a continuous length is raised and put into place all at one time.

Many homes were not originally equipped with gutters and downspouts, nor have been since. On such homes, there is frequently a water-diversion block or metal strip which prevents water running down at a house entrance. This must be removed and replaced when re-roofing the home.

Whether or not you actually need a roof drainage system depends on the amount of overhang and the condition of the ground around the home. The U.S. Dept. of HUD's Minimum Property Standards calls for guttering if the roof overhangs are less than 12 inches for a one-story home and 24 inches for a two-story. Or, if the home is located in an area having excessive soil erosion or expansion factors.

Furthermore, roof drainage conductors should not empty onto a lower roof portion, and you should try to keep all guttering clean of debris, branches, leaves and other obstructions that will allow water to accumulate in the gutter and perhaps work its way partly up the roofing edges. Allow roof ice to melt its way off rather than attempting to remove eave or valley accumulations.

A word or two about maintenance after your new roof has been completed. Avoid walking on the roof as much as possible. Most roofing materials are not designed to be the surfacing for walking decks. Keep branches from nearby trees from rubbing or scuffing the roof material. Try to anticipate falling dead limbs by cutting them down rather that having them fall on your roof.

Scope of Re-siding and Exterior Renewal

Many times a homeowner becomes dissatisfied with his home's exterior, either with its appearance or with the amount of maintenance put into it—or should be, but is not, put into it. So he suddenly decides to do something. And he jumps for the phone book, hunting numbers of home-improvement contractors and asking them over to take a look. This quick action frequently leads to the signing of a contract before the owner really knows the conditions of the various exterior parts of his home and what really ought to be done about these conditions.

You should approach exterior reconditioning with an investigatory attitude. Discover the conditions. Find out what the various remedies may be. Decide what related work can be done at the same time to create a better home, and what other improvements could logically be planned to precede or follow the prime remedies. Finally, decide what portions of the work you may wish to do yourself.

It has been seen in connection with re-roofing work that supplemental work is sometimes necessary, sometimes merely desirable, and can involve items of relatively small extra expense, such as new roof edging, replacement of certain flashings, tuck-pointing of chimneys, and perhaps the installation of new roof gutters.

With re-siding work, the related or supplemental jobs run to greater expense and more complicated work such as new soffit and fascia surfacings, a complete new roof drainage system, new trim and capping of window and door frames—or perhaps window replacements or modifications, additions to insulation, and attic ventilation.

Re-siding a home should be undertaken only after the owner has thoroughly inspected the exterior and has given due consideration to other siding-related matters. Deteriorated siding may be, and often is, accompanied by rotting eaves, nearly rusted-out gutters, and paint-peeled windows. And in these times of trying to conserve energy to keep utility bills at a minimum, there's always the question of how heating-cooling efficiency can be improved by better exterior wall insulation and proper attic venting.

This chapter, then, will help owners determine the necessary scope of their exterior renovation. Un-

derstand, though, that this does not mean the complete job needs to be done all at one time, nor does it mean that a single home improvement contractor should be the one to handle all portions of the work. Take this opportunity to have a concentrated look, and put down on paper the various things required to arrive at a logical work sequence.

An Analytic Look at Your Home

A conscientious inspection of your home's condition is not difficult. It will take some time. Arm yourself first with the tools of a housing diagnostician—a narrow paint-scraping blade, screwdriver, claw hammer and ladder long enough to reach all exterior wall areas.

Primarily, you will look for solidness and firmness. You will hope not to find softness and looseness, which indicates moisture- or rot-damage that spells potential trouble and problems. You must determine the extent of such damage, then give due consideration to tearing out damaged materials and replacing them.

As you examine the exterior walls of the home, wall by wall, you will accumulate informational details about the present siding materials, You will keep alert not just for sogginess but other characteristics such as straightness of lines, warping or bowing of present siding materials, and the condition of the caulked joints around windows and doors.

Your inspection should move on then to the roof edges, gutters, fascia boards, cornices, and soffits. These are perhaps more subject to moisture damage than any other areas of the home. This is particularly true in northern areas where some amount of snow is apt to accumulate during winter on the lower roof slopes. Check roof edges carefully for evidence that ice-damming damage has occurred. This may be in the form of separated metal edging, moisture deposits or drippings behind the fascia board, soft peelings of paint at soffit edges or joints.

Use the paint scraper and screwdriver to test for softness of wood along the roof edges. Note the frequency of soft spots and try to determine if the condition persists along all roof edges to a greater or lesser degree.

Sensible and worthwhile conversion for this 52-year old dwelling (small photo above), originally an ordinary single-family bungalow in St. Charles, Ill., the owner chose to have a complete exterior renovation job plus a conversion on the interior into two complete living units, one on each floor. To give the upper apartment its own front entrance, a window near the front corner was changed to a doorway opening to the interior stairs. On the exterior, besides new roofing, the key material was Masonite's "Colorlok" hardboard lap siding in a bright haze-gold shade. The old front porch was removed in favor of a highly appealing brick walled entrance deck.

Rear view of the renovated home indicates the addition of a pair of outdoor wood decks, upper and lower, with the latter being screened in. The upper deck for the apartment is accessible via either the outside spiral metal stairs or through the new sliding glass door in the dormer wall. Apartment's front entrance is visible at right corner. Note that upper living unit is provided with cooling through use of below-window room conditioners.

Loose nails are one small but sure indicator of poor backup lumber or framing, especially nails that have heads projecting outward and, when driven home again, do not seem to hold firmly. If many such loose nails occur, it almost certainly indicates that whatever material they're holding in place needs to be removed, and the backup lumber either replaced or supplemented by new nailing base members.

This roof edge inspection may be hampered by the roof gutters. The older strap-suspended half-round gutters are likely to reveal less roof edge and fascia damage than homes equipped with box-type gutters; the strap-suspended gutters conceal the condition of the materials behind them. It may be necessary to loosen or even remove a portion of the gutters to determine whether or not fascia boards, and perhaps lookout framing members that form the soffit base, need replacement or supplemental shoring or strengthening.

Windows, doors Close inspection around the frames of windows and doors will indicate whether or not trim work is necessary. The outer frame and trim areas are generally more subject to weathering than the sash that holds the glass. Homes older than 35 years, however, show window frame and sash problems that can be summed up by the expression "paint layering"—too many coats one on top of another, usually applied right over areas where previous layers had chipped or peeled away. Any home whose window frames and sash show widespread layering of paint is a candidate for replacement windows. With normal size, 30 inches wide by 40 inches high, double-hung window replacement units running at roughly $40 each, this can run into some money. If replacement units are the thermal type, with insulating glass, and the home has a substantial number of windows, this alone is a major expense costing more than the re-siding work.

In view of the costs, look closely at your window condition and the condition of storm windows and screens, if these are separate and in regular seasonal use. Many homeowners opt for replacement units because the changeover will eliminate the bothersome work twice each year of taking down storms and putting up screens or vice versa, plus the additional task of keeping the storms and screens clean and in good repair. More later about this aspect of exterior renovating.

Porches, stoops These areas of older homes are often more subject to the ravages of weather and time than are exterior walls. Check porch deck slopes for irregularities and inspect their roofs. Look for uneven or excessive settling. Check underdeck area for evidence of wood failures in progress. Test steps for firmness or evidence of moisture damage.

If your home has a fairly large front porch that's a bit mangy in appearance, you might: (1) completely rebuild the porch, changing it into an enclosed room for year-round use; or, (2) tear down the porch entirely and replace it with a smaller, up-to-date entrance stoop. Many, many older homes now have greatly improved appearances, even a complete "new look," because the re-siding job was preceded by porch surgery.

While considering the home's entrance areas and ways that their appearances can be improved, consider the doors themselves. Just what shape are they in? Do you feel drafts on the inside that you suspect come from the doors? Do the locks work easily and do they offer good security against break-ins? What about storm doors or combination storm-and-screen doors?

Answering these and other detailed questions will give some indication as to the need for re-

Bay window occurs where there was a narrow old window. Widening of narrow openings in exterior bearing walls generally requires addition of a supporting beam across the top of the opening, and thus a lowering of the window head height. The vinyl siding used is the "double-four" type with two four-inch bevel courses per 8-inch wide siding panel. Bay window units are available pre-assembled, in a variety of standard sizes, from millwork suppliers.

placement of entrance doors—either to function better, or to look better, or a combination of both. Investigate the features of the newer pre-engineered thermal entrance doors—units that provide excellent thermal barriers and minimal infiltration, and eliminate the need for supplemental storm or combination doors. A new entry door can go far in improving the appearance of a home.

Foundation wall Many older homes have their foundation wall tops concealed by shrubs and other plantings. Still, with increasing numbers of split-level, bi-level and walkout-basement homes that have grown older, there may be a need for treating the exposed portions of your home's foundation.

Foundation wall improvement is almost entirely

a matter of better appearance, since foundations almost never deteriorate if built of concrete or masonry. The latter, usually of concrete or cinder block units, may need attention; check for mortar cement looseness or voids that can be pointed up with fresh mortar.

If your foundation walls are for the most part visible and have an old or darkened look, you may want to help these wall areas by applying a coating of masonry paint.

Motivation Introspection is also part of the analysis of your home's exterior. This work is going to take time, trouble and money, no matter who does it. So look to your reasons for doing it and get them in order. As mentioned, the two most common reasons for exterior renovation are elimination of the annual maintenance, and an improved exterior appearance. Don't underrate appearance. A fresh clean look to a home definitely increases its value. And re-siding, plus the other renewal work associated with it, is likely to lend a measure of newness that far exceeds the better appearance usually provided by just repainting.

For many homeowners who spend more money to modernize, the main motivating element is "return on investment." People in general don't usually think of an improved home as being a generator of money or income. It can be, in two very real senses: (1) home insulating values are improved, resulting in lowered utility bills for heating and cooling; (2) with improvement of the home's neatness, appearance and up-to-dateness, there will be an increase in value and selling price.

Research for Re-Siding

If you depend upon just one or two home-improvement information sources for acquainting yourself with what materials are available for exterior renovation, you'll probably obtain a lopsided viewpoint. Those purveyors of materials or services that you contact will be pushing their own favorite product and method, discouraging any desire to learn about alternative materials or techniques.

So, to obtain a fairly accurate picture of just what materials-methods may be applicable or desirable, you should plan to spend some time in shopping around, not for bids or prices particularly, but for information on what materials are available, their performance features, relative costs, and ease or difficulty of installation.

The most common and accessible information source will be retail-merchandising lumber dealers whose business may be called something like "XYZ Home Center." Throughout the country there are also retail chains, mostly regional but a few national, operating as "home centers." These stores carry many exterior renewal materials.

A word about these home centers. A majority offer no credit, no deliveries. The stores operate on a cash-and-carry basis, although most will honor bank credit cards. But from an information-source viewpoint, you're very apt to encounter a lack of information about materials and methods on the part of the personnel of these establishments. The great majority are salesclerks without any specific experience in the use of the materials they sell.

But inquire. There will often be a "resident expert" or two on the staff, recognized by the others or by management as having a good practical knowledge of the building products the store handles. This is especially true of those home centers that have evolved from lumberyard status.

There's another angle. Many home centers now have a "literature department" offering various brochures issued by building product manufacturers, and book racks in which do-it-yourself booklets, folders, and softcover books are available.

The next research step is to use the phone book's Yellow Pages section. Look under "Siding Contractors," "Home Improvements" and "Remodeling Contractors." Each category will supplement the regular listing with display ads of the leading companies. If you phone them, you'll immediately get their sales pitches. Seldom will you be able to obtain any descriptive literature on what you're interested in. However, by checking the addresses of the listings or ads, you can probably select two or three business firms in the "Home Improvements" category that have a display room or office you can visit to inspect samples or mock-ups and pick up handout literature. In these visits you'll get more and better information by being vague about what you're planning and when you're going to do it. When salespeople smell a prospect nearly ready to do business, most will begin to pressure you with their particular brand of product or service.

With respect to re-siding materials in general, you will be able to obtain small pocket-size folders giving the products' features in brief (and often exaggerated) form. But you will seldom find, on store racks, booklets that are helpful in either product selection or in providing detailed step-by-step instructions for installing materials properly. There are many roofing-siding and home-improvement product manufacturers who regularly furnish dealers with little, condensed "do-it-yourself" folders. By their content, these are more often sales-type folders aimed at convincing you how easy the material or product is to use or apply than they are actually helpful in assisting an owner to install. Nevertheless, they're worth picking up and reading as part of your information search.

For practical, down-to-earth installation data, however, you'll have to seek out the printed instruction sheets that often come with the product itself or the installation procedures the company recommends to builders, remodelers, or application mechanics; or, in some rare cases, the detailed instruction booklets that a relatively few manufacturers have assembled especially for homeowners doing their own work. You'll find a number of such suppliers listed in the Yellow Pages along with their addresses and telephone numbers so that you can request information about their products.

Overstated Benefits

In exterior renovation materials and methods, beware of overstated benefits. This subject has been discussed at some length in an earlier chapter. But two aspects merit attention now. The first relates to the exaggerated phrases often used with many exterior renewal materials, representing them as "maintenance-free." Advertisements and literature issued by the product manufacturer often proclaim their materials to be extremely durable "lifetime" materials that you can put on your home and forget forever. The facts about the material's upkeep are seldom that simple.

The second set of overstated benefits may represent an even greater carelessness with the truth, because dollar figures are often given to indicate how much savings in fuel bills can be accrued because of the efficiency of some energy-efficient material or procedure. Energy-saving claims were just beginning to run rampant in early 1976 when the Federal Trade Commission cited a major insulation manufacturer for making false and unsubstantiated money-saving claims for its residential insulating material. The company's ads had urged homeowners to "Insulate your attic yourself and save $175 a year on air conditioning and heating." The FTC complaint citation said the manufacturer had no data to support this claim, and that the savings claims were overstated and based upon conditions which did not represent average or typical attics.

Homeowners trying to discover just where they can spend home-improvement dollars most effectively will do well to be skeptical of these exaggerations in listening to remodeling or improvement contractor sales talks, as well as when looking over manufacturers' brochures or product information folders. The discussion that follows here will provide you with at least a start in learning to cope with overstated benefits and other misrepresentations.

Annual exterior maintenance Dependable siding experts will tell you that no exterior wall is "maintenance-free." But with the newer siding materials, the factory finishes that are usually provided are of higher quality and considerably greater durability than the site-applied finishes over the wood exterior materials of yesteryear. And even where newer materials are not factory-finished, many of them come with their outer surfaces treated or stabilized in one way or another to give a more satisfactory paint base. More about these developments in a later chapter.

From a maintenance viewpoint, improved technology among the manufacturers of various kinds of building materials has reduced the need for maintenance to a fraction of what it once was for a typical solid wood, exterior painted home. Still, two of three aluminum siding salesmen will use the phrase "maintenance-free" in a way that gives two incorrect implications: (1) that you never need to do any maintenance work with it at all; (2) that this is an unequalled feature of aluminum siding.

So what should you believe? Well, first get acquainted with the idea of washing down your home's exterior, like washing your car. The newer siding materials, whether aluminum, vinyl, hardboard, fiberboard or even plywood, are not nearly so susceptible to damage by moisture. Greater moisture-resistance properties are built into more exterior materials, and factory-finishes provide a more cleanable base.

Every home's exterior acquires a film of dust, grit, spiders' webbing, and airborne debris. It collects, takes away the fresh look of new siding, tones down the sharpness of the colors, and gives the home a weather-beaten look. This is the process which, accompanied often by moisture problems, makes old sidings look bad. Yet the same kind of dirt-accumulating action on factory-finished sidings with their greater surface resistance will wash down rather easily with a garden hose and, for stubborn areas, the added use of an auto brush. Siding applicators usually recommend a once-a-year washdown to keep the home's exterior in almost-new shape.

There's a little more to it than that. Should the surface become marred or scratched, you also do what many auto owners do—use some touchup paint. Scratched surfaces on steel sidings bring potential trouble unless touched-up because of the possibility of rust. With other prefinished sidings, scratches may bring discoloration spots unless attended to.

The makers of aluminum siding and other aluminum exterior products, such as windows, storm doors, porch railings, and ventilating louvers, have long disapproved the maintenance-freedom of these materials. Today, many of those overstated claims have been toned down as countless homes bear evidence that exposed aluminum surfaces are subject to corrosion, pitting, oxidation, and other appearance-spoilers.

But to give credit where credit is due: aluminum has a number of highly desirable properties as a base material in building products, and the factory-finish technology has advanced in recent years to provide surface conditioning and a finish that needs only a moderate amount of regular care to keep it in excellent condition for years and years. And to perhaps a similar extent, the same may be said of the prefinishes on other exterior materials. And vinyl siding is without surface finish, the highly resistant material being "solid" throughout its thickness, including the color.

One more factor of exterior maintenance is the variation from one locality to another. For example, homes and other buildings close to ocean beaches often display the corrosive effects of salt. Similarly, homes near industrial plants often receive heavier-than-normal accumulations of dirt from smoke emissions. And in other areas, there may be a mildew problem with aluminum siding, causing black spots to appear. This usually occurs on protected surfaces such as under eaves or in porch enclosures. (Mildew spots are removable using a solution of ⅓-cup detergent, ⅔-cup tri-sodium phosphate such as Soilax, and a quart of sodium hypochlorite such as Clorox, all mixed in 3 gallons of water.)

And there is another pair of exterior finish offenders—tree sap and insecticides. Tree sap gets blown against the home leaving tiny speckles. Sprayed insecticides produce a somewhat similar collection of tiny spots, more localized in certain areas. In either case, these materials can be removed by the use of an ordinary household nonabrasive detergent.

With aluminum or other types of siding materials, a certain amount of color change with age is normal. In some cases, this may be due to a process called "chalking." Certain types of finishes, particularly light-colored or white ones, gradually let loose very fine particles of pigment in the form of a light powder. At one time, such chalking was explained as being not a harmful action but a self-cleaning feature to help eliminate surface dirt during a normal rainfall. Not harmful, probably true. But self-cleaning? Not true. Cleaning is something that implies uniformity and improved appearance. Anyone who has seen a chalked surface spoiling the appearance of a home can testify that this claim was probably dreamed up in an advertising session.

Insulation-related products For years and years the manufacturers and merchandisers of home products that contribute to the overall insulating efficiency of a home have shouted both verbally and in print about their products being a worthwhile investment due to the fuel savings that would occur. And relatively few homeowners paid very much attention.

Then, in the early 1970s the energy crisis was brought forcefully home to the public at large, by lines and limitations at the auto service stations, by rations in the use of home heating fuels, plus the curtailment of fuel service to new homes. And it quickly became ranting-and-raving time among the producers of insulation-oriented building products. Some insulation advertising and sales pitches implied you were something less than a patriot if you didn't help your country out of its energy troubles by immediately installing sixteen times more insulation in your home than originally came with it.

The same sort of irresponsibility in promotional claims occurred, perhaps with somewhat less frequency, in connection with other home improvement products, such as storm windows, re-siding materials, and attic ventilators. The advertising claims about the benefits of such insulation-related products had substance. The principle of saving money by adding insulative efficiency is a sound principle and can be proven. But until the previously mentioned Federal Trade Commission citation of a major company gave warning to all others in the energy-saving residential field, the exaggerated selling claims were numerous.

One example, in 1973 a booklet was issued by a vague and otherwise unidentified group calling themselves the "Ad Hoc Committee on Fuel Conservation." The booklet gave some information about the use of household energy and ways that energy savings could be made. Some purported facts and figures, minus supporting evidence, were given. One was a declaration that a typical storm window added to a home would produce a savings of from 13 to 18 percent of its purchase price in annual fuel cost. Specified for such a savings rate, the sole requirement was for the home to be located where the winter temperature averaged 45 degrees or less.

This kind of blanket statement that so many dollars can be saved irrespective of the kind of house, how well it's built, and other fuel-cost-influencing factors, is a misrepresentation.

Because added insulative value may be supplemental to roofing and siding improvements being planned, this following discussion of insulation methods is appropriate. And the subject grows in importance as fuel prices increase, and constant inflation becomes a way of life.

Energy-Efficient Exteriors

Recently, the American Society of Heating, Refrigerating and Air Conditioning Engineers (ASHRAE) issued a complex set of guidelines known as Standard 90-75, Energy Conservation in New Buildings. It is certain to become the principal policy reference for officials of many kinds in granting permission for construction of new homes, stores, shopping centers, offices, and many other types of new buildings.

This new set of design criteria makes frequent reference to a building's "envelope." The "envelope" can be described quite simply as the outer shell: roof-ceiling, exterior walls, foundation, and floor. This concept is useful in thinking about the various residential improvements that can be made to contribute to a home's energy efficiency.

Wall-ceiling-floor insulation Some outer areas of a home lose more heat than others. Or gain heat from outdoors when interior cooling is being considered. Heated air tends to flow upward. Thus ceiling and/or roofs tend to have greater heat losses per square foot than floors. Thus it is logical to provide as much ceiling or roof-ceiling insulation as practicable. Exterior walls above the ground suffer greater heat losses than floors or walls below ground level but not as much as is lost through roof-ceilings. One special note on exterior walls: the windows or glass areas and entrance doors not only suffer heat transmission losses through the material to the outside, but are also subject to infiltration losses around the perimeters. For home insulating efficiency, floors are apt to need insulation only when occurring over unheated crawl or unexcavated areas, or along slab edges where the home has a concrete floor resting on the ground. In brief, that's a sum-up of varying insulation needs. Now let's explore a few more details of providing more insulation value in roof-ceiling areas—work sometimes done in conjunction with roof improvements. Then, some considerations relating to exterior wall insulation additives in conjunction with re-siding work.

The first question about roof-ceiling insulation is, where should it be placed? In general, the answer is "just above" the living space. Now, an explanation of what is meant by "just above." Most ranch homes are one-story homes having gabled or hipped roofs with relatively low slopes. The attic space is never used for living purposes but is sometimes a place for storage. Insulation in such homes is placed in the spaces between ceiling joists just above the main floor ceiling material. In expansion-attic or Cape Cod-type homes with steeper roof slopes, the bedrooms, baths, and other-purpose rooms are finished out in the attic space. Low-height walls called "knee-walls" occur parallel with roof slopes and the ceiling is partly sloping along the rafter lines and partly flat, the finish being applied to ceiling joists or collar framing ties. In such cases, insulation is placed in ceiling-floor joists just above the main floor's ceiling and running from exterior wall to knee-wall. It is also placed between knee-wall studs and between ceiling joists just over the second-floor ceiling. And to make the insulating envelope continuous, insulation is placed between rafters before wallboard is applied to that area of sloping ceiling. In split-level and two-story homes, insulation normally goes between ceiling joists or lower chords of roof trusses. The only time insulation is placed between rafters is when the rafters or part of them serve as the nailing base for ceiling finish. In connection with the application of this type of insulation-between-rafters, it is desirable to keep a continuous air space of at least 1½ inch depth between the top of the insulation and the bottom of the roof sheathing material for proper circulation of air.

Thus, improving insulation in roof-ceiling areas is nearly always a matter of adding more to what is there now, usually in the ceiling. Insulative materials need not remain in the depth of space between joists. When this joist space has insulation to the top of the joists, additional material can be added running perpendicular to the joists and resting on their top edges.

Most insulation materials are fibrous wool-like materials available in either loose fill form, or in batts or blankets. The latter are much the same, batts being smaller flat pieces, blankets usually coming in longer lengths but rolled for handling. The batts and blankets come in two widths to fit either 16- or 24-inch spacing of joists. And they are available in varying thicknesses. All insulative materials of this type now conform to a standard guide for insulating value. The key property is the amount of heat-transfer resistance the material has or its R-value. R-values are labelled on the material and given in the insulation's literature. The common values for most fibrous or wool materials are:

R-11: 3½ inches thick for 2 x 4 stud framing;
R-13: slightly denser for 2 x 4 stud framing;
R-19: 5½ inches thick for 2 x 6 joists;
R-22: 6½ inches thick for 2 x 8 joists.

Doubled layers of insulation, of course, double the value of the resistance or give additive values if different thicknesses are used.

Homeowners may benefit from added insulation in different ways. In northern states, heating costs will be reduced; in the south, cooling costs can be lowered. In many areas, an owner will enjoy more comfort, winter and summer, at reduced energy usage and cost.

Adding to wall insulation of existing homes is not as simple a proposition as providing more insulation in or on ceilings. If your home's exterior walls now have no insulating material between studs, it was probably built before World War II. In the following period through the 1950s, it was common practice among new home builders to install blanket or batt insulation of just an inch or two in thickness. It wasn't until the mid-1960s that many new homes were provided with full-thick insulation between wall studs.

One way that add-on insulation within stud-spaces of an existing home can be provided is by the blow-in method. This technique has been in use for many years. A special blow-in type of insulating wool or fibrous material is delivered from a truck through a long 4- or 6-inch flexible air hose, through holes drilled in the exterior siding at each stud space; usually the holes are up under the eaves and are later covered. In the past few years, a somewhat similar procedure has been developed for the use of plastic foam materials as a filler of stud spaces. These highly specialized methods are mostly limited by the location and job range of application contractors to larger cities and suburbs where there are substantial numbers of old, uninsulated homes.

Insulative re-sidings Considerably easier and less expensive than fill procedures are the several methods-materials that can add insulation value over the existing siding of a home. Where re-siding requires the old surface to be furred out in order to give the new siding a solid base, urethane or styrene foam boards can be applied between furring strips. And the siding material itself may have reasonably good thermal resistance.

With aluminum, steel, and solid vinyl siding materials, an insulative backer board, fibrous, or foam, may be factory-adhered to the back of the siding; or it may come loose with the siding material for insertion as the siding is installed. Though only ⅜ inch thick, such backer boards of the newer foam materials can be quite effective. Just how effective has been demonstrated in a recent study conducted by the Architectural Aluminum Manufacturers Association. Appropriate care was used in assembling the information, and the presentation provided easy-to-understand explanation of technical details. Since a very high portion, if not all, of the insulative value in aluminum siding is derived from the material's backer board, the other horizontal sidings of steel and vinyl with similar backer boards will have comparable insulative values.

Storm or replacement windows These windows are important in connection with the new emphasis on energy conservation in homes, both existing and new. But with the volume, content, and frequency of messages sent out to builders and homeowners about insulation, one might feel that these wooly fibrous batts and blankets alone could bring a house up to a good energy-efficient level. But it's not so. Or perhaps to be more accurate, it should be said that it's only partly true. What is apparent in the marketplace is that the producers and sellers of storm and replacement windows tend to be smaller companies operating only in certain regions, and they don't have as much money to spend in communicating their energy-saving message to the public.

Evidence of the value of storm or insulating windows is ample. Two citations worth mention here are of fairly recent vintage: a 1973 study done for the U.S. Department of Housing and Urban Development calculated that for a typical 4-bedroom home in the Washington-Baltimore area, approximately $53 might be saved in the annual cost of gas for heating as the result of a $430 investment in storm windows and doors; investigations by the Citizen's Advisory Committee on Environmental Quality have indicated well-fitted storm windows/doors can save from $48 to $126 on annual fuel bills of homes in areas where the average daily temperature is 45 degrees or lower during winter months.

More down to earth in many respects than the foregoing studies is information that comes from the Small Homes Council at the University of Illinois, an organization that for many years has issued authoritative yet understandable bulletins on residential topics. On the storm window topic, the university researchers have stated that if you take ten windows without storm sash, each approximately 15 square feet in area, these units will allow enough heat to escape during winter months so that roughly 100 gallons more fuel oil will be used than for a home with storm windows. As a general rule, the SHC said that insulating windows (either storm sash or windows having double-paned glass) might produce a fuel savings of 20 percent or more, depending on the home's location and amounts of insulation and glass area.

As suggested by the preceding sentence, the amount of fuel bill savings possible for any homeowner who improves his exterior insulating envelope depends on the climate in which his home is located, the particular terrain or geographic features of his location, and the degree of insulative value his home already has. These variables make it difficult for anyone to make accurate estimates of what savings are possible. However, some guidance is provided by the following : additions to the insulating values of a home will be greater when:

• the R-values or resistance added is greater;
• the home's existing insulative value is low;

- the cost of fuel is high and still rising;
- relatively more insulative value is added to the more critical heat loss/gain areas of the roof-ceiling and windows-doors.

Heat losses and gains affect winter heating and summer cooling respectively. With windows, the losses/gains occur in two ways: through the glass (the amount depends on the glass area); through the cracks around window frames and sash (the amount being greater for longer window-sash perimeters). Thus homes with relatively large glass areas have considerably more losses/gains than do homes with small windows.

Advice given to homeowners in typical consumer articles in newspapers and shelter-type magazines for improving a home's insulative value focus on such relatively low-cost or easy-to-do items as the weather stripping and caulking around doors and windows. And as an ultimate, the adding of storm windows if these are not already part of the home's equipment (as they have been for years in many northern states).

But the plain fact is that many homes have worn-out windows and doors, and the underlying wood is coated with layer-upon-layer of paint. Or the dimensions of sash units and doors has shrunk leaving increased perimeter gaps. Latches and locks no longer assure a good fit when closed. And glazing has become loosened with frequent gaps in the contact between glazing compound and the glass. In short, the windows and doors have gone beyond their useful life, and even a major overhaul would not change them into good barriers with high thermal and infiltration resistance.

The adding of storm windows on homes whose basic window units are close to or beyond the wearing-out point is not going to help very much in improving insulative values. The owners of such homes should investigate replacement windows. Until just a few years ago, a window was a window, and there were no such things as window units designed to make window replacement relatively easy. The entire opening had to be ripped out, framing members modified and new standard-sized windows installed.

Today, there are available replacement windows of extruded aluminum and other materials which are produced on a custom-fit basis to eliminate the need for structural changes in the wall. No enlarged opening, no modification of wall framing, no need to take off interior finish or exterior siding to install a new window unit. The replacement window can often be fitted without even touching the inside trim.

Attic ventilation The proper airing of attic spaces relates both to improving a home's insulative envelope and to the application of re-siding on a home or to the installation of new soffit material. This relationship needs a chain-type explanation: What occurs in the attic of a well-insulated home is different in summer than what occurs in winter; in summer, with the more direct sunrays bearing down on the roof, attic temperatures may hit 150°F during the course of the day. Despite the well-insulated ceiling, this hot attic will transmit more heat to the interior of the home and cause longer operation of air cooling equipment if the home is so equipped. Good natural circulation of outside air through the home's ventilation openings helps to keep temperature in the attic down somewhat. Power roof or attic ventilators can reduce the attic temperature even more.

In winter, a different situation applies. The insulated ceiling allows the attic temperature to stay pretty close to the outside air temperature. But the home's living areas are generators of moisture vapor in the interior air. It comes from showering, bathing, dish washing, clothes washing, and cooking. At times, despite the air-drying out process resulting from furnace operation, the interior airborne moisture can be considered potentially dangerous. The airborne moisture vapor moves from warmer to cooler areas. To stop such movement of moisture into construction framing where it can be damaging to wood and other materials, new home building codes and HUD-FHA Minimum Property Standards require the use of a vapor barrier on the interior or warm side of all insulated walls, ceilings, and floors. Batt and blanket insulations are available with such a barrier on one side.

But even in insulated homes having such a barrier, some moisture vapor can find its way into attic spaces. There, with the lower temperatures, the vapor condenses on rafters, roof boards, and sometimes on ceiling joists. Wood members that stay moist begin to decay. Insulation that stays wet has its resistance value diminished. If you have an older home not provided with a vapor barrier and add insulation, the potential for moisture damage is greater.

But there's a solution—an easy one, in most cases. And that is to make certain your attic space is adequately ventilated with respect to the area of vent openings and their positioning for air flow. The HUD Minimum Property Standards (MPS) specify that for each 150 square feet of attic floor area, there should be one square foot of "free area" in vent openings to the outside. "Free area" is the size of the ventilator or screening and louver edges. Manufacturers of vent units, grilles, and louvers usually give the free area along with their size listing.

The Siding Specialist

Who is he? He's apt to be owner and applicator, one who works on every job himself. He's also cost estimator, purchasing agent and salesman. His office is in his home, so you can only catch him on the phone in early morning or evening hours or weekends ... unless he's then out inspecting and figuring a prospect's job.

He's not likely to offer you any financing arrangement, and if you ask, he'll probably refer you to your own bank or savings and loan association. However, he will occasionally permit customers on larger jobs to stretch their payments to him over a few months.

The typical re-siding specialist, as described above, can be found operating in cities of all sizes, practically any kind of suburb, small towns, and most rural areas. He usually tries hard to do good work. He wants no call-backs, no complaints. If you're going to have a contractor do your re-siding, re-soffit or re-guttering work, he's your man!

This isn't to say that you cannot get good re-siding work done by carpenter contractors, home remodelers and home-improvement firms. But the re-siding specialists are a relatively new breed; their work involves only the application of the newer siding materials: aluminum, steel, and vinyl. And these are the siding materials that have come to the fore in recent years, the ones that the majority of homeowners now turn to when deciding upon a complete job of exterior renewal.

Material-Oriented Businessmen

In the past, the common or prevalent home siding material was solid natural wood, sawn into boards. Beveled, clapboard and lap sidings were the usual names applied to these siding boards; the names refer to the boards' cross-sectional shapes or types of joints.

Since about 1950, there has been a very large growth in the use of wood-based or wood-processed materials developed as sidings suitable for use on residential exteriors. Plywood, fiberboards and hardboards are the three major types. Like siding boards, these processed siding materials have been distributed through normal lumber-selling channels, with retail lumberyards furnishing materials to builders and homeowners.

The marketing of aluminum, steel and vinyl siding is considerably different. Their manufacture also began in the early 1950s; increasing amounts of sheet and coil stock were supplied to small fabricating shops that were bending-forming-cutting the aluminum into siding panels and the accessory shapes which would help do a complete re-siding job. Often the fabricating shops did their own selling and installation work, or farmed the jobs out to a small applicator mechanic. Nearly all jobs were on old homes, and very little of the aluminum fabricators' products were sold to builders of new homes.

So much for history. The foregoing processes involved in developing both wood-based and non-wood products have prevailed and become predominant. Aluminum was followed by steel, and both began developing better and more durable prefinishes for the siding materials. Hardboard, fiberboard and plywood siding producers saw the competitive light, and began by priming their products for site painting, and then by providing complete factory finishes. The metal sidings found acceptance in the new home construction market. And the latest product that has made a strong leap into this field is extruded vinyl, whose rapid growth has now brought the material into price competition with aluminum.

But the separateness of the businesses has remained and, in some ways, become sharper. Aluminum-steel-vinyl go from raw material producer to automated fabricator to wholesaler and finally applicator specialist. The traditional wood-based materials flow from lumber source to processing mill to lumber broker-wholesaler and to retail lumberyard or home center. Two different channels of distribution. The wood products route deals with carpenters, builders, remodelers and do-it-yourself homeowners; the newer materials have only the applicator-specialist as a customer and contact-point with the homeowner. There must be something right about the applicator-specialist path, because aluminum has become the heavily predominant product in the overall re-siding field, although it is now being challenged by vinyl.

This digest of re-siding trends is given to help explain the reasons for dividing this text into certain chapters: in this chapter the applicator-specialist and his aluminum-steel-vinyl materials; in the next chapter, the carpenter and remodeler and his materials ... plywood, hardboards and fiberboards ... and wood boards; then, the focus is on which of the above, and what other materials, are suitable for use by the homeowner who does his own exterior renovation.

Sidings

Metal Factory-finished aluminum and steel siding materials are almost identical from a layman's viewpoint. They're alike visually because both have prefinishes front and back. The strip or panel sizes are alike, as are the various installation accessories.

Partly for this reason and partly because aluminum out-sells steel by at least 10 or 12 to one, this text will focus on aluminum siding. Readers should infer that virtually everything said about aluminum can also be said about steel sidings, and where differences do occur, they will be mentioned.

Plastics Vinyls are not entirely new as a siding material. B.F. Goodrich Chemical Company has been one of the pioneering companies in developing rigid vinyl building products. But what is new about them is the rapidly broadening distribution and acceptance by applicator specialists, and the stabilizing of prices to become competitive with aluminum. Another relatively new element on vinyl sidings-soffits-rainware is the entry of major building product manufacturing firms into this vinyl siding field: Celotex, Bird & Son, Johns-Manville and GAF Corporation, to name a few.

Vinyl siding and trim are much like aluminum siding in sizes, shapes, types of accessories and application procedures. There are some differences in characteristics and some in application, which will be pointed out. One main difference from aluminum is the vinyl surface. There is no applied finish. The vinyl is homogeneous throughout its thickness. For this reason it is often called solid vinyl, or rigid vinyl, to distinguish it from many vinyl materials that are flexible and clothlike.

In seeking information about siding materials, you'll quickly become aware of the benefits of the various types of products, but the negative points are not likely to be mentioned. As with almost any type of building product, each type has a shortcoming or two. Aluminum siding and steel, to a lesser extent, are subject to damage both in application and, subsequently, by denting on impact or due to careless handling. Their finishes can be scratched or marred rather easily. With steel, there is the added concern of damaged spots becoming prone to rust. Also, with metal sidings, there is a need to provide a grounding path for protection during electrical storms. Applicators find vinyl materials become brittle when outdoor temperatures run below 45 degrees Fahrenheit. Cracks that can occur during cold weather application could become points of leakage.

Aluminum siding Aluminum, through the years in various forms of building products, has gained a reputation as a low-maintenance material. Basically, it is a noncorrosive metal. But because the plain metal surface is subject to pitting, oxidation or other appearance-detracting action, aluminum products are given factory finishes, often quite sophisticated, well-researched finishes. With sidings and accessories, durable face finishes may be in the form of baked-enamel coatings or ultra-thin plastic film coverings. Back or concealed sides of aluminum panels or trim are given lesser-resistant coatings simply to eliminate the oily film usually on the surface of raw aluminum, thus helping minimize fingerprint and handling marks.

The baked-on finish is generally a two-coat process applied over both the smooth and embossed sidings. The embossing process, incidentally, is to provide a woodgrain texture to the surface and is done with aluminum and vinyl materials. Single and sometimes dual film coatings provide better corrosion resistance and more uniformity to the finish, resulting in better weatherability, abrasion resistance and finish integrity. You pay more, of course, for the better finish.

Factory-finishes on metal, as well as those on various wood-based products discussed in a later chapter, are subject to a certain dulling with age. The siding manufacturers are probably correct in telling you their siding products will endure intense sunlight, snow-sleet, salt-spray and other wind-blown items. But siding colors do change. With time, they lose the pureness of color-tone they had when new. They acquire a film of grime that's sometimes difficult to remove. While there's no danger of wearing or rotting out, their surfaces get a jaded look and need a freshening coat of paint.

There are variances in metal thicknesses used in aluminum sidings. This is not of great concern because nearly all fabricators of siding panels and trim adhere to the practice of using thicker 0.024 gauge metal for panels or for shapes in which the flat portions are wider than 5 inches and without any backing material. For shapes having unbacked portions under 5 inches wide and for panels with backing for wider areas, the normal thickness is 0.019 metal.

The backing mentioned above is in the form of an insulative material of one kind or another. Aluminum siding without backer board has negligible insulating properties. Nor is it much help in reducing street noise heard within the home, a claim sometimes made by over-enthusiastic siding salesmen.

Is it fireproof? Being a metal, yes. It won't burn, but it does not materially add to the fire-resistance of the wall. On the other hand, certain exterior finishes might ignite and spread if the siding is subjected to very intense heat, but the burning of the finish on the outside of a home is not of significant danger to the home's occupants.

Aluminum siding panels come essentially in two types: a beveled strip having an exposed 8-inch wide face for horizontal application and a flat or slightly-curved panel of similar width designed for vertical application on exterior walls. A variation in the beveled type is a panel or strip having two 4-inch wide bevels for a narrow clapboardlike appearance.

Embossing of the surface to give a wood-grain look is done lengthwise. A few companies also offer a rustic-looking surface with embossing in the form of curving saw marks running perpendicular to the panel length. All panels of different fabricating firms run between 12 and 13 feet in length.

In addition, many fabricators offer horizontal bevel panels in which small ventilating louvers have been formed. In vertical siding panels, which do double-duty as soffit material, perforated panels are available for venting use.

Color selection in aluminum sidings may or may not be very wide depending upon the fabricator and also depending on the range of stock carried by the wholesaler. One large fabricating manufacturer shows his product line in the form of a color guide chart which indicates that 13 different types of sidings are offered in a choice of 17 colors, mostly light-toned pastels, and inside-outside corner posts, undersill trim, L- and J-channel trim and barge moldings can all be obtained in matching colors.

There is another type of aluminum exterior product, briefly mentioned in an earlier roofing chapter. Aluminum shingles, or shingle-shakes as they're sometimes called, are being produced by only a few firms, but these are major companies (Alcoa, Reynolds, Kaiser). Originally marketed for roofing work, these shingle-shake panels are becoming popular for use with mansardlike roof installations and for exterior siding,

Higher in cost than conventional aluminum siding, the shingle-shakes as a re-siding material probably pose fewer problems in application and thereby may be suitable for some homeowners desiring to do their own re-siding.

Solid vinyl As you observe for the first time a strip or panel of vinyl siding being applied to a home's exterior, you are apt to think "perfection." Though the panels are not fully rigid—they have some flexibility—their application is accomplished to avoid stress on the material after it is in place. The result is that the siding edges and horizontal lines appear to be unusually straight and in excellent alignment. The colors, somehow, seem very pure. The embossed surfaces have almost a sculptured look. The uniformity, panel to panel to trim, is striking.

Vinyl's properties are probably about the equal of aluminum with respect to maintenance and upkeep. It remains to be seen from actual on-the-house experience what vinyl's durability and appearance continuity will be. Application specialists who have worked with both aluminum and vinyl say the vinyl is an easier material to work with and not subject to the little accidents that can spoil an aluminum panel. Whether or not the through-penetration of color will be of particular value still is in question, as well as just how long the material will retain its purity of color.

As with aluminum, vinyl siding can be cleaned of normal dirt and grime accumulations just through washing with mild household detergents to remove more stubborn stains. The surface of vinyl is such that peeling, flaking or chipping is unlikely. Vinyl is impervious to any mildew action or rusting; it does not dent; it is paintable, should painting be desired some time in the future.

One shortcoming of vinyl is applicable only in certain areas of the country. The material does become brittle in cold temperatures. Applicators say that when it does become brittle, there is a possibility of cracks developing from impacts. It may be appropriate at this point, with denting and impact damage to aluminum and vinyl sidings being discussed, to state that such accidents are more "repairable" than with most other kinds of sidings. Individual panels of aluminum and vinyl siding can be removed rather easily, using a special hooked tool to unlock the material interlocking edges. A damaged panel is removed and replaced with a new panel.

In cost, vinyl was higher priced than aluminum when first introduced, but increased usage and greater distribution has brought its cost to the point where the two materials are price-competitive in nearly all areas of the country. And though the various vinyl shapes and accessories available are nearly identical to those in aluminum, the choice of vinyl colors offered by most fabricating companies is limited to about a half a dozen—white and the more popular pastels. But vinyl's popularity is spreading fast, and you will no doubt soon see at least a few firms offering sharper color tones, such as gold, dark green, brown and blue, as the vinyl formulations improve. Like aluminum, vinyl is available in an embossed wood-grain surface as well as smooth, but the embossing is slightly deeper and thus, in appearance, more pronounced.

Vinyl is a more natural material for carpenters to work with than the metal sidings. Builders of new homes are finding that carpenter applicators can

Renewed with aluminum siding, graphically evident in the before and after views of this old one-story cottage, which had reached the decrepit stage. A siding special-

ist cannot just re-side but also will retrim the windows, roof edges, porch columns and entrances. In short, he will do a complete exterior renewal job.

become accustomed to working with vinyl in less than a day. Vinyl has not yet proven itself as a do-it-yourself material, but this is almost sure to come, in this writer's opinion. It handles easily. The usual carpenter tools are quite appropriate for cutting and applying. Only two extra hand tools are used for special purposes: aviation-type curved-blade tinsnips for fine trimming and mitering and snap-lock punch tool for turning a fresh-cut horizontal panel's top edge into a crimped and interlocking edge.

One characteristic of vinyl is its dimensional change with temperature. Wood expands and contracts minimally with temperature but substantially with moisture content. All metals expand-contract slightly with temperature. Vinyl moves a little bit more, and the installer accommodates this movement in the nailing method. All vinyl panels have nailing slots. Nails are positioned near the centers of the slots and are not driven up tight. The result is that the vinyl siding "hangs" on the nails without stress or binding to the under surfaces. This allows freedom in expansion-contraction movements.

Vinyl siding panels are normally 12½ feet long. The horizontal bevel types of siding have an 8-inch exposed width or a pair of double 4-inch bevels. The vertical panels have 8-inch exposed faces whose surface is slightly concaved. The vertical panels also serve as soffit panels. The typical installation accessories and trim pieces included by most fabricators are:

starter strips	inside/outside corners
J-channel trim	soffit frieze-runner
undersill trim	fascia cover

A few fabricating-manufacturers can furnish special window-doorframe covering strips, batten strips, and similar special-purpose covering materials.

Cost Expectations

As a general rule (that has known exceptions) aluminum and vinyl re-sidings provide economical exterior renewal of most homes in comparison with other re-siding materials. Asphalt and asbestos sidings are outdated although their low prices may make them appear worthwhile for some do-it-yourself homeowners. The trouble is that such low-priced materials add to the homeowner's work without adding to the home's value.

In respect to costs that re-siding application specialists quote to homeowners, it's believed that these prices will be consistent, having minimal spread from one specialist to another. While inflation has, in recent years, yielded a steady rise in both materials and labor, and is almost certain to continue to in years to come, here is some idea of the rule-of-thumb costs that re-siding specialists are quoting in the early 1980s.

Briefly, the homeowner can expect to pay approximately between $125 and $200 per hundred square feet (known in roofing and siding circles as a "square") for either aluminum or vinyl re-siding applied by a specialist. Soffit-fascia renewal is currently quoted in the range of about $5 to $10 per linear foot, the lower figure for narrow soffits only a foot or less in width, the larger figure for wide soffits or overhangs of 30 to 36 inches. Steel sidings, with a higher material cost to the applicator, are apt to be quoted at prices 30-to-40 percent more than aluminum or vinyl. There can be variances in these price levels, also, in different parts of the country due to different shipping costs and labor rates.

Wood Siding and Carpenter-Remodeler Exteriors

As indicated earlier, the traditional channel of distribution for lumber and wood products has had the retail lumber dealer and carpenter or remodeling contractor as the final links.

In this chapter, the various materials these companies can offer the homeowner will be treated in some detail. Readers should not infer from this chapter and the previous one that a carpenter or home-improvement contractor would not be able to quote you on a re-siding job with aluminum or vinyl. There is nothing especially exclusive about sources of supply for the siding specialist or the remodeling contractor. Each prefers to follow his own class of materials on most jobs. Some siding specialists will occasionally apply other materials with which they're most versed. And the probability is that you'll get both better prices and better workmanship if you have your home re-sided with the materials they're experienced in handling.

The range of work capabilities of the carpenter-remodeler is likely to be considerably greater than those of the average re-siding specialist. With a background in building trades, and most cases considerable experience in general carpentry or building of new homes, the carpenter-remodeler is usually qualified to discuss with you the possibility of making changes that might affect the house structure, whereas the re-siding specialist is rarely knowledgeable in this respect.

Making House Modifications

The suggestions that homeowners considering re-siding or exterior renewal commit some details to paper was stated earlier. The purpose for getting such details and possible remodeling ideas down in black-and-white becomes evident now, when you are ready to call out remodeler contractors or home-improvement firms to look at your house. If you have given suitable thought to the subject and put down on paper what you'd like to accomplish in addition to the re-siding work, you'll be better prepared to talk, ask questions about possibilities, and communicate your needs or desires in order to obtain suggestions.

There is another good reason for preparing yourself to discuss alteration or remodeling details. The contractor must first discover what you have in mind in order to determine the initial step—drawing up suitable plans that can be submitted in application for a building or remodeling permit. In most localities, simple resurfacing of a building does not require a permit, whereas any modifications or additions to the building structure ... or modifications or additions to its plumbing, heating, electrical systems ... do require a permit. And this applies whether the work is to be done by a contractor or by the homeowner.

Very often, the carpenter or remodeling contractor is capable of furnishing sketches or drawings that are sufficient for permit purposes; this varies with locality and is more likely in small town and rural areas. Cities and suburbs frequently require that drawings and plans for house alterations carry an architect's or engineer's stamp indicative of at least a check-over by a qualified structural professional. And in many localities, you may also be required to obtain a permit for any wrecking or tearing-down work.

Windows One area of the exterior envelope most vunerable to moisture or deterioration damage is the windows. As mentioned earlier, loose and poorly fitted old windows are a source of major heat losses/gains. More than that, as many new buyers of older homes have discovered, poorly fitted windows cause drafts in the room, often to the considerable discomfort of room occupants. As a result, a carpenter or remodeler is often asked "what can be done" about the windows. Well, replacements are one answer and there are a growing number of window manufacturing firms able to supply nearly-exact-size replacement units. But some homeowners are desirous of enlarging certain windows, or grouping units in a certain wall, or providing an outward-projecting bay window.

Porches Another common source of deterioration trouble is the porches once commonly provided at each home entrance. Porches often rest on post foundations, are frequently the areas of insect attack, and have settling unevenness that becomes obvious in sagging deck and roof lines. Carpenter-remodelers are capable of suggesting methods of either porch repair or replacement.

Overhangs Old siding may show evidence of

moisture damage resulting from condensation within the wall as a result of the lack of a warm-side vapor barrier. But the siding may also show evidence of moisture damage due to exterior run-down and trapping of rain or melting snow. This occurs frequently with homes having little or no roof overhangs. Consideration should be given to extending overhangs if new roofing is part of the exterior renewal work. Old decorative moldings and trim frames may also be victims of trapped moisture, and replacement is often the only solution.

Choices in Re-siding Materials

Carpenters and remodeling contractors tend to be traditionalists who prefer using the older materials. If a home has horizontal lap or clapboard siding, you'll virtually never receive a contractor's suggestion that the home's appearance might be improved if all or part of the old siding were covered with the newer sheet or vertical panel materials. So, to acquaint homeowners with the wide range of wood exterior materials now available, the balance of this chapter will emphasize the products available, their benefits and possible shortcomings.

After you've become interested in the siding possibilities of certain products, don't depend upon questions put to carpenter-remodelers for details. First, visit retail supply or home center stores to obtain literature on the types of products attractive to you, then discuss them with the contractors you call in to give estimates.

Before presenting details on siding materials, however, there are points to be noted about siding trends that may affect your material choice. Once, many older homes were commonly re-sided using asphaltic board panels with mineral granules pressed into the board's surface in bricklike or stonelike patterns. Though low in cost, asphalt siding is obsolete for residential use though it may still be available in some areas and used for recovering sheds, barns and other out-buildings.

The situation is a little different with asbestos-cement sidings. At one time, there were many manufacturers of such sidings and the material, though subject to easy chipping and breakage, became quite popular for a long period in the 1940s and 1950s. In today's context, any product using the word "asbestos" could encounter merchandising problems. And while there may be one or two companies still producing what was commonly called asbestos siding, this type is also virtually obsolete.

Also gradually becoming part of the past are numerous kinds of solid wood or natural wood sidings. At one time, it was common for sawmills all around the country to turn out siding boards in various regional wood species. These were commonly in the drop, shiplap and bevel siding shapes. That practice has gradually changed as wood sidings lost the competitive battle on new home construction to other new siding materials that cost less and apply faster. In many cases, wood siding materials cut in flat-grain style gave way to edge-grain sidings which have proved superior in their paint-holding properties. Yet another factor in the siding trend picture has been the rise in preference for prefinished sidings among new home builders, and remodelers as well.

This preference for prefinished siding materials has broadened as new and better factory-finishes have been developed. And this growth in sophisticated prefinishes, begun by metal and hardboard siding producers, has spread to plywood and fiberboard materials. And the latest development in conjunction with prefinishes is surface texturing of one kind or another to simulate grain, sawing marks, or other details.

Hardboard and fiberboard Wood fibers ground down to almost pulp form, then processed and pressed into flat sheets, have become widely used building products under the names "hardboard" for thinner, denser materials, and "fiberboard" for somewhat thicker and lighter-density sheets.

Some of these wood-fiber board or sheet materials have been around the home-building field for a long time and to many readers, the products may be more familiar by their trade names. The pioneering firm in hardboards was the Masonite Corporation, whose hardboards were characterized by the smooth surface on one side and the screenlike cross-hatching on the reverse side of the 4 x 8 foot sheets. The brown-colored product was available in either of two types—untempered and tempered, the latter being harder and denser.

In the light-density fiberboard sheet material category were those called insulating sheathings. In 4 x 8 foot square-edged sheets or 2 x 8 foot tongue-groove panels, insulating sheathing was pioneered, logically, by a Minnesota-based firm called the Insulite Company. Later, Insulite developed a medium-density fiberboard siding material which is still marketed by the Insulite Division of the Boise-Cascade Corporation.

Also among the early-comers in fiberboard products were two eastern manufacturers, the Upson Company and the Homasote Company. Their products became known simply as Upson-board and Homasote-board and these names were probably better known in small town and rural areas than they were in urban locations. These firms are still producing fiberboard products on a much broader scale. Once used primarily as interior wallboard, water-resistant formulations have been perfected and various forms of these products are now avail-

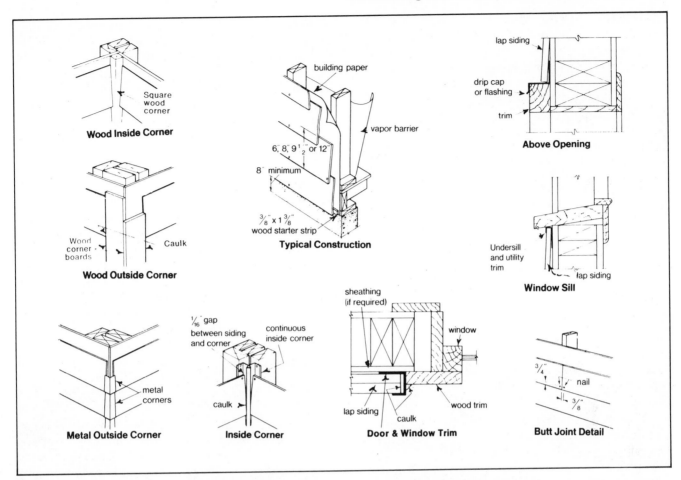

Corners and joints of hardboard siding installations are handled in certain common ways indicated in the sketches above for lap or bevel siding and below for sheet or panel siding.

These illustration details appear in the hardboard siding catalog of Champion Bldg. Products for the firm's "Sundance" and "Cadence" siding lines, and the proce-dures will generally be applicable for hardboard sidings made by other companies as well. However, some firms producing plastic-film surfaced sidings may recommend installation in a slightly different manner. Note: sketches shown are intended for new construction applications, but joint and corner treatments when re-siding will be essentially the same.

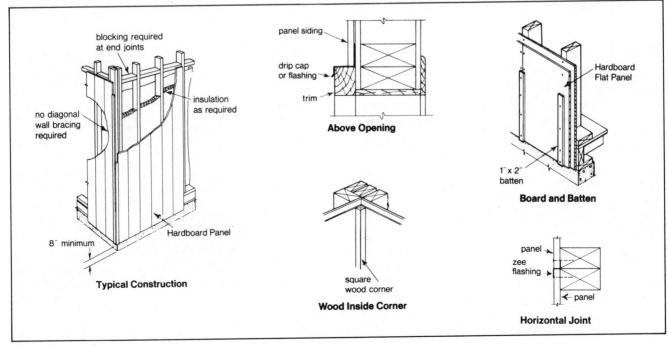

Lap Siding Panels of hardboard in 12-inch width and 16 feet long are being applied to a room addition. The siding is prefinished smooth-face siding. Other hardboard sidings in lap form include textured woodgrains and thatch or shake types. Photos courtesy of Masonite

able for exterior sheathing and prefinished siding purposes. Both are relatively inexpensive materials.

The initial hardboard sidings were lap type in 12-inch widths, and furnished plain or with a primed finish. Now hardboard sidings come in a wide variety of both lap and full 4 x 8 sheet sizes, with various textures easily produced in the production molding-pressing operation. At first textures were simply simulations of coarse-grained wood boards but later came other styles, quite realistic in their appearance, simulating rough-sawn wood and thatch-types having the look of cedar shingles and shakes. One especially realistic simulation is Masonite's light-colored "Stuccato" sheet siding which simulates coarse-troweled cement plaster surfacing. This material is being used increasingly in conjunction with stained boards or half-timbers to give the modern equivalent of English Tudor styling without problems involved with exterior wet plastering.

Lap or bevel types of hardboard siding are gener-

Start application at base of exterior wall, where applicator is tacking in nails at chalk line level for accurate placement of 1 x 3 or metal starter strip. In photo right

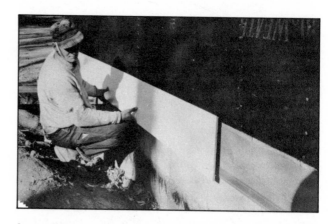

above, first course panel is placed to fit over the metal starter strip.

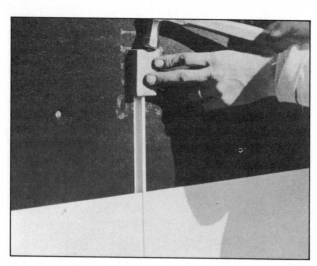

End connector supplied by some siding manufacturers is an H-shaped metal molding that slips over panel ends where they meet.

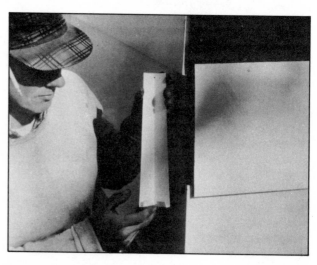

Metal siding corners are used at each course coming up to outside corners.

ally offered in 9- and 12-inch widths, panel lengths of 16 feet. Panel thickness is 7/16 inch and most styles are offered in choice of several colors of prefinish or in a prime coat for field finishing. The newer textured styles are usually offered in prefinished or prestained.

Hardboard sidings are easy to work with ordinary woodworking tools. In this respect, their one fault lies in an ability to take a sharp edge quickly off a saw blade. The use of hardboards have good dimensional stability compared to solid wood or other wood products, some contraction/expansion is possible and a gap of 1/16 inch is suggested where hardboard joints butt. Where siding panels butt, metal H-shape joint moldings are available that provide for expansion and contraction movements.

The application of 4 x 8 and 4 x 9 prefinished hardboard siding sheets with vertical patterns is particularly easy on either new construction or re-siding jobs. Styles available include several types and spacings of grooves as well as separate or integral battens. The grooving and battens, of course, are utilized to conceal sheet joints and also to keep nailheads from becoming too apparent. Hardboard siding sheets are usually 3/8 inch thick although some with deeper grooving or integral battens are apt to have the 7/16 inch thickness.

Hardboard siding manufacturers supply a full range of accessory moldings and shapes. These generally include starter strips, color-matched butt-joint moldings, metal outside-inside corners, J-trim for window-door frames, Z-flashing for horizontal butt joints and batten strips with prefinished metal snap-on covers. In addition, many hardboard manufacturers can furnish complete fascia-soffit materials.

Relatively new among the fiberboard sheet materials suitable for exterior siding and soffit uses are certain overlay materials. For example, one such product has been named "Cladwood" and is described as a "homogeneous lightweight medium density-board with a resin-impregnated fiber overlay on both sides." The material is said not to raise or check, nor will its surface crack or delaminate. Another somewhat similar product is "Durasote" whose finish is a 3 mil thick solid-color acrylic film factory-applied by waterproof adhesive to an extra-rigid insulative fiberboard.

The plywood group Like hardboard, plywood siding is available either in horizontal strip or lap form and in vertical pattern sheets. Also like hardboards, the newer plywood sidings are textured.

Among the many desirable properties of plywood as an exterior siding material are its uniformity in surface, one sheet to another, and its structural reliability. Plywood sidings have good rigidity, excellent impact resistance, and long durability with relatively minimal maintenance. They are, in addition,

This re-siding job uses 3/8-inch textured plywood plus some 1 x 3 battens. Also added was a new entry door and a front yard privacy screen.

1. *Nailing the siding is normally done with 6-penny box galvanized or siding nails for material half-inch thick or less. In this re-siding case over stucco, hardened nails are needed to penetrate stucco. (Photos in this series are courtesy of the American Plywood Association).*

economical.

All plywood sidings employ waterproof glues in their laminations. Prefinish practice varies. Some plywood manufacturers offer a wider range of prefinished plywood sidings than others. But unfinished sidings are readily available and, to a lesser extent, primed sidings.

While some plywood mills specialize in certain types of plywood materials, and other major marketing firms selling plywood vary in the types of plywood sidings handled, on a national basis there are three principal types of siding materials available.

- Standard—"303 Sidings" refer to grade marking by this number, and are characterized by a wide choice of textures, most of which are intended for use with stain finishes:
- Medium Density Overlaid—commonly called MDO, these sidings have a special resin-treated surface hot-bonded to the plywood panel to give it a uniformly smooth base for paint;
- Special Coatings—sidings provided with special surface finishes, including one type with pebbles or stone aggregate adhered to the plywood.

Horizontal plywood lap sidings are generally 3/8 inch thick, in widths of 6, 8, 12, or 16 inches and in lengths of 8, 12, or 16 feet. Some companies provide slightly beveled top and bottom edges, assuring a drip edge whichever way the panel is placed, thus minimizing the waste when siding is cut on low-slope angles for gable ends.

It is in vertical-pattern sheet sidings, however, that a really broad variety of surface textures, prestains, and prefinishes are offered. Sheet sizes are 4 x 8, 4 x 9, and 4 x 10. The oldest patterning device is grooving. Different spacings of grooves plus differ-

2. Plywood sheet siding covers a lot of wall area in the 4 x 8 foot size. Shown here is full-size sheet with window cut out. A slight ¼ to ½ inch gap is left around the window for easier fit.

ent widths and depths of grooves create a wide range of patterns. Most grooved styles have shiplap edges coordinated with the grooving so that there is no evidence of where sheets join after the material has been applied.

Flat ungrooved sheets with various texturings are also available and in application these frequently have 1 x 2 or 1 x 3 batten strips applied at regular intervals. The joints between sheets are part of the interval incrementing thereby concealing and sealing the joints. Some plywood siding manufacturers can supply plywood batten strips having a matching texture and prestain.

Plywood siding thickness varies between ⅜ and ⅝ inch depending upon the nature and depth of the grooving. Groove spacings may be equal on the same sheet, as in the popular "Texture 1-11" style sidings, or the grooving may be irregularly spaced in order to give the appearance of random width "boards" between grooves.

In re-siding work with plywood lap or sheet sidings, the surface preparation work described earlier to obtain evenness and alignment of the old surface panel is important. Any bowing or depression in the old siding is very likely to be passed along and can be evident in the new siding surface unless corrective measures are taken before applying the new siding material.

There's one factor involved with vertically pat-

3. At the garage opening, the textured plywood siding is applied over the door (short waste pieces used) and, as indicated in photo right, over the entire face of the garage door. Garage door hardware may need adjustment to compensate for added weight of siding and battens.

4. Batten strips of plywood are applied at regular intervals (approximately 16 inches), but their spacing is adjusted to fit between window distances; battens occur at sides of windows and doors. Smaller trim molds are used above and below windows to cover gaps.

terned plywood sheet sidings to which homeowners must give due consideration. That is the possible change in appearance of a home. The horizontal lines of traditional bevel or clapboard sidings tend to give most homes a closer-to-the-ground look. If such homes are re-sided with vertical-pattern sheets, the vertical lines will result in a taller appearance and may, in some cases, conflict with the general architectural lines of the home.

One final aspect of plywood sidings: they're fine materials for use by the do-it-yourself homeowner. And more will be said on this in the next chapter.

Solid, Natural Wood

The long-traditional exterior wood materials remain favorites of many homeowners, remodelers, and builders. This is especially true in certain regions of the country, such as the Pacific Northwest, California, the Mid-South and parts of the Northeast where lumbering has long flourished.

But use of solid wood sidings and wood shingles, or shakes in other areas of the country, has been overtaken by the increased use of faster-applying materials, particularly the factory-finished materials. This appears to be true for both new home sidings and re-siding jobs.

An underlying reason for these use changes is, of course, the relative costs of the materials. Natural wood exterior materials range near the top in square-foot cost, being exceeded in most areas only by masonry or masonrylike materials. Despite this, the natural woods have been found acceptable for new homes in the moderate-to-higher price ranges and for commercial construction of all kinds. Wood sidings and wood shingles-shakes are not about to disappear from the marketplace.

The proponents of wood exterior materials, and thus the best source of information about wood products, are trade associations rather than individual and competitive companies or producing mills. These associations also serve the builder, remodeler, and consumer by the work they do in conjunction with mill inspections and grade-marking of lumber, sidings, and shingles. Associations involved with exterior siding materials can be located through most roofing and siding dealers.

The lumber industry as a whole is a complex one, serving many fields in addition to home building and remodeling. While the industry does maintain certain statistics reporting data flow information relating to general product categories, such as dimension lumber and boards, there is, to this writer's knowledge, no source of information about the popularity rise or decline in the different kinds of siding shapes, grades, or species.

Few people will question the continuing popularity of horizontal bevel siding. And though for a time

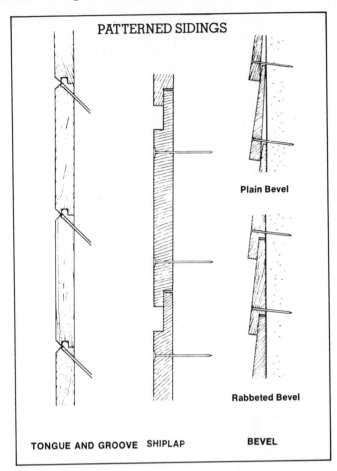

Solid wood sidings in various wood species are still sold by many lumber suppliers, but principally the bevel type is available. In redwood, however, a wide range of patterns are available and, for a still further choice of exterior appearance, square-edged siding boards in redwood can be used as indicated above. Sketches courtesy of California Redwood Association

the wider 8- and 10-inch bevel types appeared to dominate, recent years have seen a return to the narrow-width Colonial clapboard styles containing two or three bevels per board. At one time, too, some mills or distributors began offering siding with a prime coat and, to a limited extent, this practice continues.

Finishes Time has indicated flat-grain siding boards (the kind whose grain pattern is now being simulated by other materials) are not very capable in their paint-holding characteristics. Today, most homeowners desiring wood as a re-siding material will be more satisfied if they pay slightly more in order to obtain edge-grain siding boards—unless they plan to purchase rough-sawn boards which will receive a stain finish. Knotty woods are not desirable as sidings regardless of finish, because, over time, shrinkage will probably cause drop-out of the knotty cores.

Much of the difficulty encountered on older homes in keeping up finishes and appearance has stemmed not so much from the wood itself as from a moisture condition. Most frequently, the source of the trouble is the lack of a suitable vapor barrier on the interior side of exterior walls. Such a lack often causes, in cold weather, a process of condensation within the wall and the passage of moisture vapor to the outer portions of the wall. And zing, there goes the paint.

As a rule, most sidings should be painted or stained to prevent exterior wall faces from excessive wetting due to rain, snow and dew. Exceptions occur with certain species such as cedar and redwood, which are sometimes left unfinished because of their ability to weather attractively. Keep in mind that an ability to weather to a nice hue is accompanied by an ability to stain badly at the drop of a paint brush, the drip of tree sap, or the splatter of roof sediment being cleaned out of gutters.

Often a very satisfactory solution to the natural look is a transparent finish—like varnish but better than varnish. That is, better than the old linseed-oil varnishes which inevitably turned yellow. There are some excellent transparent finishes available, including urethane and acrylic varnishes. There are also the penetrating preservative-type finishes that stabilize the natural wood's appearance while making it more resistant to the effects of moisture.

For painting of new sidings, prime and topcoats using acrylic latex paints are recommended. These are particularly easy for the homeowner to apply. Many homeowners will also find a very satisfactory alternative in pigmented stains. The heavy-bodied type of stain is appropriate for use with smooth or rough siding boards where covering or hiding power is desired. The light-bodied stains are appropriate where it is desired to color-tone the exterior but still have wood grain or texture partly show through.

Installation Wood siding boards pose few installation problems, although nailing practices do vary somewhat according to the siding shape and joint details. Where possible, siding courses should be adjusted slightly in spacing to run continuously above and below windows, and above doors, to minimize the need for notching. Overlaps should run 1 to 1½ inches, the greater amount for wider boards.

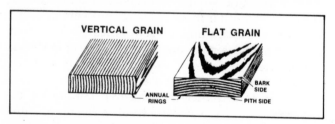

Vertical grain has greater stability than flat grain, and weathers exceptionally well. It is usually preferred for exterior finish carpentry. Flat grain redwood is highly figured as compared to the more subdued uniform pattern of vertical grain redwood.

Unlike metal and plastic siding materials, wood boards should be cut squarely and butt tightly to window and door casings or frames, and to building corner trim pieces. The old practice of mitering siding boards at corners has given way to vertical corner trim pieces or to individual metal corner units. Use of these metal corners is no longer considered, as it once was, a cheap practice. It is laborsaving and does present a neat corner appearance, although some siding applicators claim that the nails that hold these corners in place loosen over time.

Wood shingles and the thicker and naturally split shakes, once thought to be primarily roofing materials, are enjoying a steadily increasing acceptance as exterior siding materials. Virtually all shingles-shakes are of cedar, and while there are a few mill

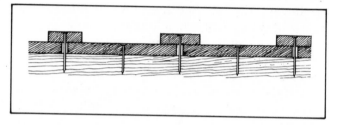

Standard board and batten: drive one 8d nail midway between the edges of the underboard at each bearing. Then apply batten strips and nail with one 10d nail at each bearing so that shank passes through space between underboards.

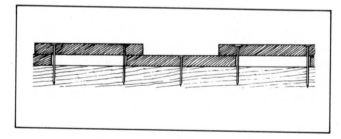

Board on board: space underboards to allow 1½-inch overlap by outer boards at both edges. Use one 8d nail per bearing for underboards. Outer boards must be nailed twice per bearing to insure proper fastening; use 10d nails, driven so that the shanks clear the underboard by approximately ¼ inch.

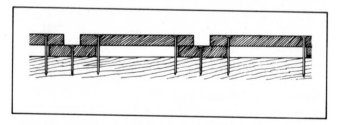

Reverse batten: nailing is similar to board on board. Drive one 8d nail per bearing through center of under strip, and two 10d nails per bearing through outer boards.

sources that supply northern cedar shingles, the bulk of the materials are derived from western red cedar by Pacific Northwest mills.

Among uses of wood shingles and shakes, there's an unpredictable trend which appears to have two facets to it: (1) architects and design professionals of all kinds are finding that shingles and shakes have great flexibility for innovative applications on interior walls as well as exteriors; (2) a revival is occurring in the use of "fancy-butt" shingles, long ago the mark of accent decor on exteriors of Victorian-styled houses.

Fancy-butts are being seen more often though you may not have known their name. They are uniform-width shingles, usually 5 inches wide by 15 inches long. Their exposed butt ends are trimmed at the mill into various shapes. Half-rounds and V-points were popular in olden times. Now, some shingle producers are offering them in nearly a dozen different shapes: round, diagonal, square, arrow, acorn, hexagonal, octangular, fish-scale, diamond, and half-cove.

Wood shingles and shakes, despite a fairly high cost, have great popularity and acceptance. They can be stain-finished in a variety of ways. Some say exterior shingles or shakes can remain unfinished and weather to a beautiful gray. Truth may often be otherwise; the weathering process does not occur uniformly. And unfinished shingles or shakes are subject to water run-off stains. Recommendation: carefully investigate stains and preservatives available, and select a finish that will preserve the material's excellent appearance and also resist the susceptibility of unfinished cedar to stains.

Wood shingles and shakes, like some of the materials described later, are excellent do-it-yourself products that give much enjoyment in their application as well as saving in the labor costs of home improvements. Yet, being a wood product, shingles-shakes are also materials with which carpenters and many home remodelers are familiar.

Regional Siding Preferences

What has been said about exterior siding materials, their acceptability status among homeowners, their suitability for re-siding homes, their availability among local stocking suppliers and the scope of their usage, must be somewhat tempered by regional preferences, practices, and economics. Easy examples: redwood siding is a beautiful product used extensively on the West Coast but difficult to find among stocking dealers in eastern states. Southern pine materials are seldom seen outside the South.

The same applies with other lesser-used materials. Cement stucco and plaster are seldom used as a re-siding or new siding material outside of California, Arizona, and Florida. Asbestos siding always sold better in the northeast and midwest.

To some extent these regional preferences still prevail and perhaps it is the applicator specialists, builders, remodelers, and home improvement firms that unconsciously foster the continuance. For these businessmen, each improvement job is different, yet their business requires them to estimate its cost in advance although there are many contingencies that may erase an expected profit. In contrast, a high proportion of new home builders build-and-sell only after costs are pretty well pinpointed and contingencies limited. As a result, the re-siding contractors and remodelers are prone to stick to familiar materials and do not quickly jump into using new or innovative products. And for somewhat similar reasons, they are to some degree more wary of lower-priced materials, and alert to better quality products that carry good backup warranties or guarantees.

Doing Re-Siding Yourself

There are some homes that are very difficult to re-side. There are many houses of pre-1900 vintage that would make basket cases out of many a skilled and experienced re-siding specialist. Such houses are the tall, multi-dormer and multiwinged ones of 2, 2½, 3, and even 4 stories, the latter often being on the lower or rear side of a hillside location.

Some re-siding materials are also more difficult to work with than others. For example, cement plaster and stucco refinished work is certainly beyond the capabilities of over 90 percent of the homeowners who consider themselves pretty fair at do-it-yourself arts and crafts. Plaster and stucco work take some practice. You don't want the evidence of your unskilled practicing anywhere on the outside of your home, even the rear.

There's another factor, too, that differs home to home, and that is the condition of the material that is presently on the outside of the home. If the house has settled irregularly, if it has some obviously rotted spots that could be early surface warnings of a structure underneath that needs repair, or if the exterior walls have numerous bulges and curving lines—these all spell trouble and the ambitious do-it-yourselfer will be better off bringing in professional men to deal with such cases.

Still another area of predecision concern is the matter of availability of materials and equipment. Materials differ in how easy they are to get, how long it takes to get delivery, or how far you have to go to obtain them or pick up additional amounts or forgotten items. And with certain materials, metal sidings for example, a good job of re-siding can only be done through the use of certain special tools and equipment. Small hand tools pose no problem; it's the items that have to be borrowed or rented—portable brake, pneumatic stapler and compressor, scaffold equipment—for which you must give advance thought to the time needed for you to do the job. Will you be working on it every day, or just on weekends? In view of the rented equipment cost, will you be saving enough to make the whole do-it-yourself project worthwhile?

Time and Trouble

Make a time and material estimate. This first step is aimed at: (1) determining the approximate amount of re-siding materials and accessories that you will need to do the job, so that when you visit suppliers you can obtain price information and quickly convert to an approximate total the material cost; (2) determining the approximate amount of application time.

Calculating square footage and time The effort to work up your estimate is simpler than you might suspect. You start by calculating the area of each exterior wall that the siding will cover. On each wall measure full wall width. Use a 10- or 12-foot long 1 x 2, 1 x 3, or 2 x 2 for obtaining wall heights. Mark it off in six-inch increments, and with a ladder, use the measuring pole to get two-story and gable heights. Also, when measuring obtain the height and width of each window to the nearest quarter-foot (3 inches).

The wall area, or course, is the number of square feet you get multiplying wall width by wall height, and subtracting the window and door area on that wall. For gable ends, measure the height down from the roof peak to the wall height at the eaves. Multiply gable height by wall width and divide by two for gable-end area, then add this to previously figured wall area. Keep notes on each window or door size. With dormers, measure and figure face wall as you do an exterior wall and gable; measure dormer eave length and multiply by dormer height and you have the area of both dormer sides.

Now, with your measurements complete and noted down, it will be relatively easy to obtain a figure for the overall total area that the re-siding material must cover. How much material is needed to cover that will depend on the material itself and what is lost in laps. This conversion can be easily calculated when you talk with the supplier. You will also add two waste percentage allowances: (1) general waste in cutting or trimming to fit, and in end-lapping or joining—usually 10 percent; (2) roof-slope waste on gable ends and dormers—from about 15 percent for steep slopes to perhaps 40 percent for very low-pitched roof slopes. In connection with point (2) above, it should be noted that a few re-siding materials in hardboard or plywood are "reversible." That is, a siding strip may be applied with either edge up. Such a provision will save on angle-cut waste.

In addition to re-siding area, your notations on windows, doors and exterior wall heights will also make it easy for you to estimate the quantity of trim or accessory items needed. And your building width figures will assist in determining amount of starter-

Hand tools for siding application when using aluminum (shown at right) are suggested in the Installation Manual for applications of Alcoa Building Products. Practices and tools used by siding specialists may vary from one area to another. The tools shown are identified as follows: (1) chalk line reel, (2) folding rule, (3) 2 or 4 foot level, (4) carpenter's steel square, (5) caulking gun, (6) electric saw with aluminum cutting blade, (7) claw hammer, (8) double-action aviation snips, (9) utility knife, (10) metal file, (11) fine-tooth hacksaw, (12) tin-snips, (13) conventional cross-cut handsaw, (14) nail set.

stripping. Wall and window-door dimensions will also give you the basis for calculating the quantity of furring strips required, if this form of surface preparation is to be used.

These various calculations should precede your decision as to whether you will handle the re-siding work yourself or have a contractor do it for you. By making such a time and material estimate, you will be able to compare the approximate costs of the two methods. And, as you consider the type of siding material you desire to use, you will obtain an idea of the relative speed of application. Having the wall dimensions at hand in organized fashion from this time and material estimate, you will then be able to make a quick estimate of what the overall time of your application work will be. Since it is human nature to overestimate the amount of work that you can do in a given period of time, do this:

- add 15 percent labor time to your original estimate if your plans call for the use of large-sized or long materials that don't need much cutting;
- add 25 percent to your original estimated labor time with materials that need to be cut to short lengths or special cuts such as miters;
- add 20 percent to your original estimated labor time when materials have to be applied from near the tops of ladders or scaffolds above one story in height;
- add 10 percent to your original estimated labor time if your helper(s) is a non-handyman type or a family member.

Also, another contingency in estimating time of application is the weather. Hot humid weather will slow you down. Rain or snow, if light or intermittent, will slow the work and perhaps be messy enough to cause some extra cleaning work. But your biggest potential enemy on the weather front is wind. Strong winds will hamper application work of nearly all kinds. It is particularly difficult to handle larger materials in a strong wind. And certain materials, notably aluminum or steel sidings, can be damaged or ruined in a split-second by a strong wind gust. Not to mention the hazard of accidental damage to applicator and helper(s).

Availability of equipment Let's assume a

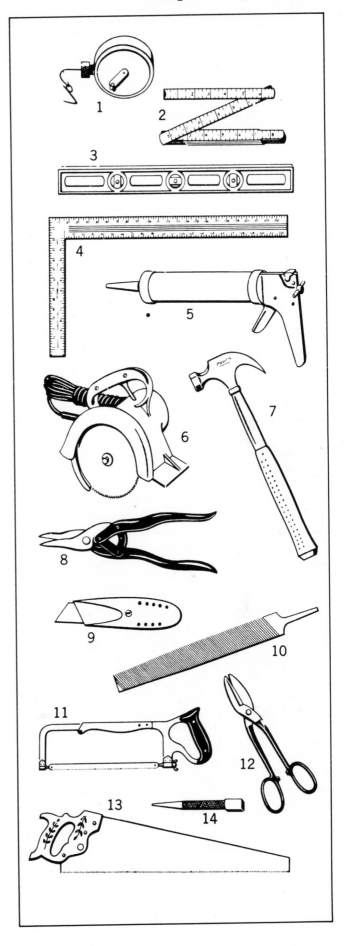

homeowner deciding to do his own re-siding work is willing to spend some cash or credit card money for some decent tools. What is meant by "decent" is simply "professional." Home fix-up tools are now available in drug stores, grocery supermarkets and discount outlets. In most cases, the tools such stores carry are cheap, price-competitive imitations of the kinds of tools that tradesmen use. Don't buy them. Go to a well-established hardware store, a contractor supply house or a retail lumber or home center. Most of these suppliers will have several grades or price levels in each kind of tool. Buy either the top or next-to-top grade. It will, in nearly all cases, help you do the work better but, more importantly, it will help avoid frustrations. You'll be far less exasperated working with decent tools.

For doing your own re-siding work with any wood, wood-based or wood-fibrous product, a virtual necessity is an electric handsaw. A portable electric circular saw in the 7-to-8 inch blade size may cost between $30 and about $90. Forget any saw under about $60. The problems with cheap saws are twofold: they not only have insufficient power to keep blades from binding and slowing in many materials, but their adjustment devices for precision and accuracy do not enable precise and accurate cutting.

In conjunction with your electric handsaw, plan on spending a fair chunk of change for preparing a decent worktable. At minimum, this would involve devices for making four sawhorses and two or three 12-inch wide, 2-inch thick planks as a work top. Preferably, the work top should be 18 to 20 feet in length if horizontal strip or lap materials are to be used. If sheet size materials are to be applied, the work top may be shorter but wider, perhaps about 5 x 12 or 5 x 14 feet. With strip or lap material, in which the most common cut will be right-angle cutoffs, it would be advisable to either buy or build a

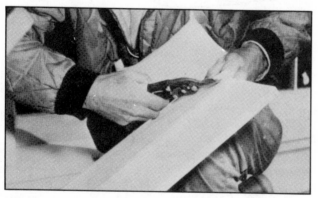

Cut and crimp tools for use with vinyl siding include the precision aviation-type curved-blade snips whose double-acting blades easily follow cutting marks in vinyl and a pliers-like hand tool called a "Snap-lock Punch." It makes a depression crimp in vinyl cut edges that allow the edge to be inserted into trim strips and locked into position.

cutting jig for your electric handsaw to run in.

Now, about the larger items, or more expensive ones that may not be within your purchase range. These may or may not be available from local rental agencies. If not, then you should inquire of re-siding specialists or contractors about the possibility of renting these kinds of units from them. They may have older extra equipment that they don't use regularly any more. They may be able to make certain items available to you for limited periods, such as over weekends. While this equipment may not be absolutely necessary for every job (depending upon the house and the re-siding material to be used), these tools can save much time and effort:

Radial saw. An alternative to an electric handsaw, but a superior tool because it is mounted on its own table for which it is easy to provide extensions; more importantly, it allows more accurate cutting and easier cutting of precise angles.

Scaffolding. Two common forms of siding work are: (1) aluminum telescoping planks that provide a working platform when ends are supported by ladders or scaffold wall brackets; (2) foot-operated scaffold jacks that ride on vertical 4 x 4 or 4 x 6 upright posts—wood or aluminum planks stretch across the jacks, which can be raised or lowered by foot-levers.

Metal brake. A portable device that can rest on sawhorses and provides a means of uniformly bending metal right angles. The brake is most commonly needed for work with metal sidings, and is virtually a necessity when the work involves the capping or covering of window/door frame trim or custom fitting of coil stock to cover rake, frieze or fascia trim.

The partial-contract concept Some of the foregoing comments no doubt are discouraging to would-be do-it-yourself re-siders. Maybe the estimates showed the re-siding job would be too time-consuming, or might stretch over too long a period. Or, maybe inquiries made indicated an inability to rent or borrow suitable equipment. In such cases it may be possible to work out a compromise—a help-share-the-work compromise. You should approach it in the following manner.

After you've arrived at a preliminary decision on the type of re-siding material you feel you're most interested in, and after you've decided on any of the supplementary exterior jobs that you might wish to proceed with, then begin calling in suitable specialists, contractors or home remodelers to give you prices. Ask them to separate their estimates or prices according to the different jobs because you're not certain yet as to how much of the work you wish to have done all at once. Get one bid on the siding with or without window capping or retrimming. Get a separate bid on soffit-fascia work, on replacement windows, on new roof drainage system.

Obtain firm proposals in writing that give you this separate price breakdown. Obtain such proposals from four or five re-siding firms if you're in an urban or suburban location, two or three if in a rural location. In making the contacts, observe the nature of the men contacted. You'll be able to judge those who sincerely desire to provide you with exactly what you want.

It is just such a sincere and conscientious contractor that is most likely to do a complete and satisfactory job. But it is also the same sort of eager-to-please contractor who might be willing to split the work with you in one way or another. And you can put it to him openly and on the line—you're trying to save every nickel you can on the job. Would it be possible for him to take on the job and have you act as his helper for certain portions of the work, at a reduced price?

Such a proposition could fall upon listening ears. If you're catching him in his busiest season, he may be working short-handed and your offer to work with him might be the difference that would mean his taking the job and his not being able to do it until some later date. And if you contact him in his dull season when some of his regular helpers have been laid off, your offer might mean he wouldn't have to hire an additional man.

There are other ways to split work. He'll do the main house; you do the garage. He'll do the window and wall work; you do the eave-and-fascia recovering. And so on. Talk it over with him. You'll be surprised at how accommodating some re-siding and remodeling contractors can be. But you must be prepared. You must show him that you've checked into different materials. That you know what needs to be done. That you know you can do a good job with his guidance. And so on. He isn't inclined to leap into such arrangements without giving them due thought. And in preparing the contract agreement you will want to have him spell out exactly what work he will do, what work you are to do, and what the price adjustment will be.

Material-Related Problems

Some re-siding materials handle and apply easier than others. Just what the characteristics of the various materials are will be dealt with briefly and an idea of the kind of difficulties that may be encountered will also be given below. The reader shouldn't infer from the negative comments on certain products that he should therefore avoid them completely. The plain fact is that in building and remodeling work, almost every material or product that you use is capable of misuse in one way or another. All that's being given here is a forewarning.

Aluminum, steel and vinyl The metal siding

Saw-horse and plank scaffolds are quite suitable when working on the siding or soffits of a one-story home. A scaffold using three or four 2 x 10s or 2 x 12s will allow several men to work in the same area with relative safety.

Use care in attempting to suspend scaffold platforms from ladders. Before doing so, investigate availability in contractor supply firms of special suspension brackets that provide proper attachment to ladders.

When using ladders, either for working purposes or simply to reach scaffolds, always face the ladder while going up or down. Lean toward the ladder when working, and try to keep one hand in a holding position or free to make a quick grab.

materials are subject to relatively easy damage in handling, in cutting, in moving to the wall position and in nailing. The most noticeable difference between the siding specialist and the carpenter-remodeler is the concentration on what each is doing. The specialist keeps the metal sidings out in front of him when he carries or handles the pieces. He is very alert to the locations of the ends of the pieces and he looks before he swings the panels or strips around. In cutting, the specialist gets down close to the saw blade to see that it enters the material at exactly the right place. The carpenter-remodeler is often a bit more casual, knowing that a slight miscue won't make that much difference nor will there be damaged material.

Aluminum siding materials are easier to work with than steel. With steel, any slight surface damage, incorrect cutting, or accidental finish-marring spot is apt to develop into a rust spot after application. Such spots are more difficult to correct and refinish than the same type of damage on aluminum siding. However, most experienced craftsmen will find steel siding's slightly extra benefits worth the higher price and handling-application difficulties.

Aluminum, on the other hand, is also worth do-it-yourself consideration. Although nearly all manufacturers of aluminum siding believe their products beyond the capabilities of do-it-yourselfers, they are underrating the capabilities of many of today's homeowners. Because aluminum siding and accessories are distributed through different channels than most building products, the homeowner may have trouble finding a supplier of aluminum siding materials. In locations near larger cities, finding a supplier won't be difficult. A few aluminum fabricating companies are beginning to market their products through retail lumber or building supply outlets and home centers.

Because solid or rigid vinyl siding and accessories are being sold through pretty much the same channels as aluminum and steel, the homeowner desiring to buy his own vinyl and apply it is likely to run into a supplier hunt much the same as described above for aluminum. But there's this difference between aluminum and vinyl—vinyl is almost sure to prove itself as being an excellent material for do-it-yourselfers to work with.

Vinyl materials still can't be handled carelessly, but there is a big difference as far as slight mishaps are concerned. An entire vinyl panel won't be completely spoiled if just a slight wrong twist is given it. Sliding the panel accidentally over the edge of a splintery 2 x 4 or 2 x 6 won't put a nice long irrepairable scratch on the surface finish. Deburring the sawn edges of vinyl with sandpaper or file won't insert fine metallic slivers into the skin of the applicator.

Despite the above encouraging words about vinyl sidings, aluminum and vinyl must be considered as sensitive materials compared to wood or wood-based materials. The inclination here is ↑ to recommend aluminum and vinyl for homeowners having some previous do-it-yourself home product experience and to limit this recommendation to re-siding work on simpler homes of 1-, 1½-, and 2-story construction.

One very brief word about a special kind of aluminum exterior material. Aluminum shingles, sometimes called by the manufacturer shakes or shingle-shakes, were first introduced a few years ago as a roofing material. Now, a growing number of applications are being seen on exterior walls. The individual shingle or shake sizes are roughly 12 x 36 or 12 x 48 inches making for easier handling. All four edges interlock. The aluminum surface is heavily textured to resemble handsplit wood shakes. These aluminum shingle-shakes pose fewer application problems than regular aluminum siding as far as do-it-yourself application on exterior walls is concerned. But finding a supply source nearby may be even more of a problem.

Another detail to remember about metal sidings: they are materials along which lightning discharges can travel. They need to be grounded so that an easy path is provided for any such discharge, rather than take a chance that the electricity could skip around near the ground where people might be endangered. To connect metal siding to a ground conductor, an approved lug-connector is screwed to the butt edge of the siding near the grade level. A copper No. 8 size wire is then run from the connector to the home's regular electric service ground, or to a nearby cold-water pipe using an electrical pipe clamp.

Sheet siding materials These are probably easiest of all re-siding types to apply. But for many homes they pose a design problem because of their vertical patterning. The appearance of homes whose existing exterior material has horizontal lines will undergo a radical change when vertical pattern siding is used. Proportions change. What was formerly a nice ground-hugging home can become an upright building with a funny look. The wiser homeowner will take the time to sketch out elevation drawings of the front or other easily viewed sides of his house, and include the vertical patterning to approximate scale before making a final decision to use these kinds of materials.

The problem of a vertical visual pattern for re-siding is more likely to be present with homes of 1½ or more stories. Vertically grooved or battened sheets seldom look inappropriate on 1-story ranch homes. The common practice with new home designers to avoid exaggerated vertical lines on a building is to limit the area in which the vertically patterned sheets are used. Gable ends, front entry or porch walls, above/below windows and wainscoting are some of the limited areas.

Hardboards, plywoods and fiberboards offer very few problems for the do-it-yourself re-sider. Nailing practice varies and the most important aspect here is to utilize the type of nails the material manufacturer recommends, and to use special care with hardboards and fiberboards to keep nail positions back from board edges the recommended distance. The only other principal caution is for care in the handling and nailing of prefinished sheet materials. The finishes are durable to weathering once in place, but take care not to damage them by carelessness in carrying, positioning and fastening; this makes scars and scratches that take extra time to retouch or repair.

Great variety in grooved or battened patterns is available in both hardboards and plywoods. Smooth, semicoarse and rough textures are offered. In fiberboards, the surfaces may be smooth or lightly textured. Many manufacturers of these sheet materials also offer lap siding strips. These

Outside corner post is positioned on building corner with ¼-inch gap at top for expansion; nails spaced 8 to 12 inches. Metal corner units are normally not used with vinyl siding.

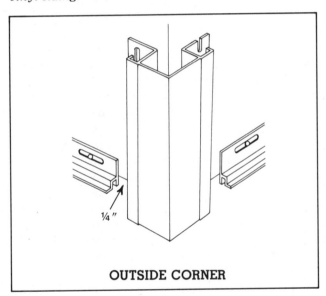

OUTSIDE CORNER

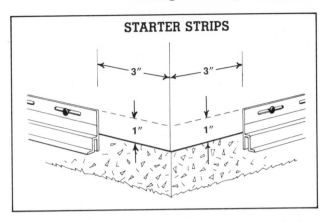

STARTER STRIPS

Starter strips, as in drawing above, are kept back 3 inches from building corners so that corner posts can extend downward to point just below bottoms of the strips.

The extruded vinyl starter strips are nailed in slot centers, not quite driving nails up tight. The strips butt to each other, again leaving a ¼-inch expansion gap.

Use of a chalkline and level is recommended around both sides of each building corner, in order to assure that the starter course is accurately level and conceals the bottom of old siding.

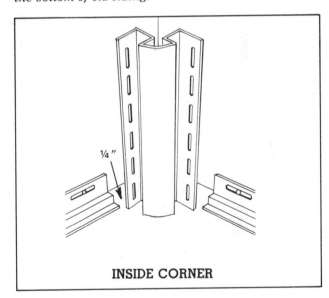

INSIDE CORNER

Inside corners apply in similar way to outside corners with gaps on each side to starter strips. Note that siding panels must be cut to sufficient length to fit within the corner post channels but not fit tightly; there should be ¼-inch gap between the panel end and the channel back. Corner posts are of extruded vinyl in matching colors.

apply much like bevel siding but usually only wider widths are offered, 12 to 16 inches. Lap sidings in these materials in some cases are offered unfinished but are available usually with a prime coat. A few manufacturers offer factory-finished lap sidings. Because of the greater likelihood that lap sidings will be painted, some manufacturers provide the material with a surface overlay that gives more uniformity and paint-receptivity.

In applying sheet materials for re-siding, keep in mind the following points, which may or may not be a factor in your work.

- The most common and readily available sheet size is 4 x 8 feet. But most producers also offer 4 x 9 and sometimes 4 x 10 sizes. In fiberboards, some manufacturers offer 8 foot wide sheets: 8 x 12, 8 x 14, 8 x 16 feet.
- Some prefinished siding producers offer match-

ing finish batten strips and these solve the problem of trying to match up the raw wood's finish.

- With most sheet materials, allow a slight expansion gap between sheets or shiplap joints, ⅛ inch being about right in most cases.
- For better appearance of prefinished siding materials, use near-matching colored nails and touch-up colored putty sticks.

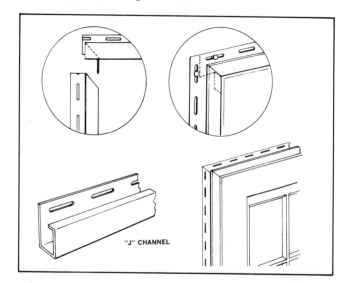

"J" CHANNEL

Window trim moldings of J-channel shape are of extruded vinyl in colors matching siding. They are placed at sides and top (jambs and head) of each window. Mitered corners with top fold-down tabs are used as indicated in sketches.

Proper nailing of vinyl siding is shown at the right. Corrosion-resistant nails are used and positioned in the centers of the nailing slots at about 16-inch spacings. Be careful not to drive nails fully home; they should be left just slightly loose so the siding hangs on the nails. This permits siding panel movement in expansion and contraction due to temperature changes.

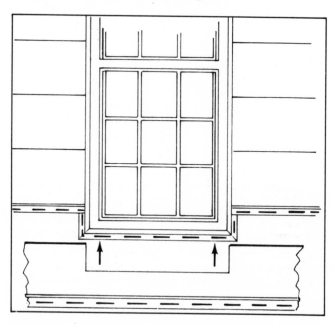

Over windows, siding panel receives a U-cutout similar to that for under windows but made from the bottom of the panel.

Be sure to measure width carefully between J-channel backs leaving ¼-inch gap at each side.

If old siding doesn't fit window frames well at head and jambs, caulk the cracks before installing the J-channel trim members.

Below-window fitting is handled by use of a vinyl undersill trim strip whose shape is shown in the sketch. Blocking will ordinarily be needed under the trim strip, unless the siding panels happen to fall, so no panel cutting is needed below the window.

Tinsnips can be used for making vertical cuts to fit sides of window. A utility knife can be used to make a horizontal cut by using a steel square as a guide, and the U-shaped under-window cut completed, with waste piece to be discarded.

Crimping tool is used horizontally along raw cut edge to place projecting nubs which will lock into the under-sill trim when the panel is slipped upward into position.

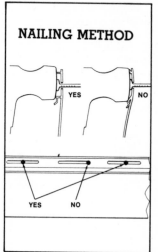

NAILING METHOD

YES   NO

YES   NO

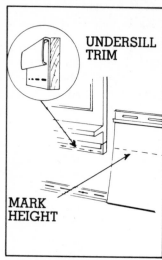

UNDERSILL TRIM

MARK HEIGHT

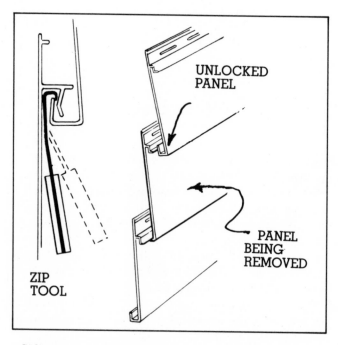

UNLOCKED PANEL

PANEL BEING REMOVED

ZIP TOOL

Siding panels progress upwards with pieces cut to proper length for fit between windows or between window and building corner. Endlaps of panels, if needed, should be staggered one course to another, and overlying lap ends should face away from entrance doors.

- Though ready-primed sidings offer a little protection against dampness or moisture, store the delivered materials inside under cover ... and still plan to apply finish coats to all primed exterior materials as soon as possible.
- For natural wood finishes using plywood sidings, redwood and cedar materials will accept a variety of clear or preservative finishes and yet keep the character of the wood and its graining visible.

Wood shingles and shakes Although these are often associated with roofing work, wood shingles and shakes make excellent sidewall materials and the average do-it-yourselfer can hardly go wrong with them.

The application can often go on directly over old sidings, but a layer of 15 pound felt building paper should be applied first. With some houses where the old siding is bowed or warped, or where the old siding is hard to nail into such as masonry or stucco, the use of wood furring strips is advisable. Use hardened masonry type nails for applying the furring, then normal shingle nails for the shingles or shakes. Furring strips should be spaced according to shingle or shake length positioning them so the strip centers are about 2 inches above the line where the shingle butts will occur.

The same shingles and shakes described earlier for roofing purposes may be used for exterior wall applications, but the amount of shingle exposure on walls may run one to three inches greater than for roofs. In addition, building products suppliers in many areas will offer a special kind of what is called "rebutted and rejointed" materials. Edges have been machine trimmed to be exactly parallel and at precise right angles with the trimmed butts. The purpose is to provide shingles with tight-fitting long edges, which, after application, will give emphasis to the horizontal butt lines. It is common practice with these sidewall shingles as well as with regular roofing shingles to "double-course" them on sidewalls in order to give a thicker butt appearance. Double-coursing simply means a double layer, and usually lower-grade and lower-priced shingles are used on the under-coursing work.

Factory-primed or factory-dipped shingles mean better color penetration and less need for extensive finishing work after application. With good factory finish, just touch-up finishing is needed on the site. With the preprimed shingles, a more uniform finish is usually possible but be sure to follow the manufacturer's finishing instructions in order to avoid a clash between his preprime and your site finish.

Especially recommended for creative types doing their own shingle work are the "fancy butts" mentioned in the previous chapter. These uniform-width shingles in 5 x 15-inch size are generally applied with a 6-inch exposure, but there is considerable freedom or flexibility in their butt designs and how they can be combined or applied. Some suppliers furnish fancy butt shingles only to custom special orders. For a long time, retail dealers did not stock them. But they're making a comeback and you may soon be seeing cartons of the more popular butt designs in lumber and home center outlets.

Solid wood Natural wood-board sidings are not as easy to find in retailer stocks as they once were. While the different types stocked may vary considerably in different regions of the country, some of the old favorites are seldom seen: the familiar drop siding with its curved configuration to accommodate the butt lap, the Dolly Varden style, and several types of tongue-and-groove boards.

On the other hand, there's increased use of plain square-edged boards, the standard boards with battens, board-on-board, and boards with reverse battens. There is also broader distribution of western siding boards, notably redwood and cedar, which offer great possiblities for clear and semitransparent on-site finishing.

While wood sidings have long been associated

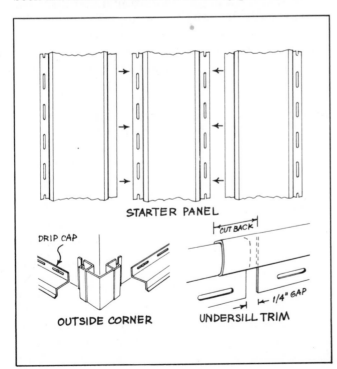

STARTER PANEL

DRIP CAP

OUTSIDE CORNER

CUT BACK

1/4" GAP

UNDERSILL TRIM

Another type of vinyl siding is that shown in the drawings at left. Wall installation of vertical siding begins with a double-sided starter panel at the center point between outside wall corners. Then, additional standard panels are slipped into the interlock edges on each side of the starter panel.

At the base of the vertical panels, a drip-cap trim member is installed instead of a starter strip. Outside and inside corner installation are the same as for horizontal siding panels. The undersill trim pieces are installed to hold the tops of vertical panels in place.

Gable and dormer endwalls sometimes pose a problem in design when re-siding a home. Don't overlook the possibility of using uniformly trimmed 15-inch shingles, 5 inches in width. These "fancy butts" are currently being offered by several shingle manufacturers and are seeing use on interior walls as well as exterior.

At left above, on the gable-end above the second floor dormer are octagonal-butt shingles. The dormer wall is finished in random-width regular wood shingles. At the right, the photo shows a second floor exterior with square butts intermixed with diamond butts, while under the gable peak are octagonal butts.

with paint peelings or other surface faults and with the need for fairly frequent repaintings, this association and reputation is not entirely deserved. A high proportion of faulty exterior paint situations is due basically to the lack of a vapor barrier on the warm side of exterior walls.

In addition, paints and other exterior coatings for wood sidings are of much better formulation than they once were and many types are better able to cope with possible moisture problems not to mention durability. Some modern paints, those based upon polyvinyl acetates for example, are apt to need repainting at only about 10-to 12-year intervals. Many acrylic latex paints will remain in excellent shape for 4 to 6 years.

Bevel or lap sidings in natural wood generally come in 1 x 6 and 1 x 8 sizes, and in double-beveled narrow 3- or 4-inch clapboard style.

Thin masonry Because brick and stone veneers rate high in homeowner acceptance and because real 4-inch brick and stone veneers require a troublesome add-on foundation wall projection, an entire new industry has been developed: the thin-masonry product industry.

The above may be an exaggeration, but only a slight one, really. The plain fact is that today you can walk into nearly any home center, hardware store, painting-decorating store or retail lumber and building supply establishment and find some kind

of product whose finished surface is designed to look like brick veneer and, to a lesser extent, stone veneer. Call them brick simulations, call them false brick, call them synthetic stones, whatever. One adjective seems to apply to nearly all of them regardless of the actual material of which they're made—thin. All of the simulating materials are of substantially less thickness and of considerably less weight than the conventional clay masonry or natural stone building materials.

Some thin brick materials are excellent products for handling and application by homeowners. Others have limitations or are more complex in application. Some are designed for interior use only and shouldn't be used on a home's exterior. On a square foot area basis, most are relatively expensive in comparison with other siding materials. To give readers a brief picture of the different kinds of materials available in this category, the following paragraphs provide a capsule description.

Roxite—premortared brick and stone panels produced in 10 x 48-inch size; brick panels have 24 molded-in brick faces per panel and weigh under five pounds per panel. The material is 60 percent real crushed limestone; it is reinforced with fiberglass, and the panels are formed under heat and pressure with polyester and acrylic resins. Natural glass and the panels are formed under heat and pressure with polyester and acrylic resins. Natural earth pigmentation gives the brick and stone faces an authentic coloring. The panels nail up like siding and there's no mortaring or leveling. The materials saw smoothly; the panels have interlocking flanges that hide nails fastening the previous panels, and corner accessory panels make neat turns around outside corners.

Panel brick—panels use real thin bricks of ½-inch thickness bonded to an asphalt-impregnated backer board that is also ½-inch thick. Panels have interlock edges and special L-shaped corner brick units providing a stretcher that is brick length on one side and a header length brick face on the other. After application of panels, spaces between brick faces are mortared either by hand or by using a special grout pump. The application is by nailing between brick units, the nails concealed by the later mortaring. Each panel is 6 brick courses high and panel sizes are either 16½ x 24 or 16½ x 48 inches. Weight is about six pounds per square foot.

Brickettes—individual brick face units ½-inch thick, available in two types: 2½ x 8 and 4 x 8, the former to provide edge facing while the latter is primarily for floor use.

Cultured stone veneer—series of different kinds of very realistic, thin synthetic stones of unusual textures and colors. The stones lay up in a conventional mortar bed when applied over masonry. Over wood, the application involves building paper, wire lath, a scratch coat and a mortar bed. The material has found rapid, widespread acceptance by contractors for apartment and commercial buildings.

Job Pointers and Procedures

The great variety of re-siding materials available and suitable for do-it-yourself application precludes detailed descriptions of step-by-step application details. However, an excellent understanding of what work is involved can be absorbed by study of the various illustration panels accompanying the text in this and other chapters.

There is always a further complication in providing detailed installation instructions. Recommended procedures with nearly identical products of the same type are quite likely to vary from one manufacturer to another. Each maker of materials has a somewhat different group of suggestions to make that he believes will insure satisfaction from the use of his product. These suggestions generally take the form of a printed instruction sheet that is included in each package of materials. Too often, a homeowner's difficulties in application are due to a disregard for, or lack of attention to, the manufacturer's instructions.

While many building product and material makers accompany the material with instructional aids, there are often some points not covered or related items that are not given due consideration. In the following paragraphs, the reader is exposed to a series of general pointers which can be of assistance in the installation of various re-sidings.

Wall preparation, layout Avoid placing new material over old, loose siding boards by firming up the old exterior material where necessary, even to the removal and replacement of damaged or deteriorated areas. Re-sidings need a solid base for nailing.

Check the alignment of old surfaces by eye noting any outward bowed areas or depressions. Attempt to eliminate such bowing or warping by saw cuts and renailing. If such methods are ineffective, apply wood furring strips over the old siding material, spacing the strips for the re-siding and using shims or wedges to obtain even alignment over irregularities in the old siding.

Focus special attention upon getting a level start along the base of the old siding. Your goal is to have a good, visually level base line when the house shows evidence of settling or when the bottom of the siding is at different levels on different sides of the house.

If re-siding is of the horizontal strip or lap type, a useful homemade measuring tool called a "story-pole" will come in handy. Normally used in masonry coursing, such a story-pole is made of a straight 2 x 2, 2 x 3, or 1 x 4 that's reasonably stiff. Select a pole about 8 or 10 feet long and begin at the bottom with a series of sharp visible marks. Each mark represents an additional course or siding strip location and the spacing will be suitable for the siding width being used. Make sure, however, that the marks give the normal siding width, less the overlap distance or the exposure width of the siding. The story-pole can be used not only to mark course positions but also to check strip alignment after siding is applied.

Windows, doors, corners Window-door frame conditions may have to be modified to accommodate re-siding materials. The key factor is the amount of projection the window-door frames and trim have outward from the face of the existing wall. The frame condition is often complicated by trim moldings, or shaped-rather-than-flat frames.

Some window-door frames can be capped by use of custom-braked aluminum stock. In other cases, shaped portions of the frame may have to be cut off and replaced by nailing on new flat or square-edged trim pieces, mitering them at corners.

The most common mistake resulting in poor window appearance is the application of narrow trim pieces so that the re-siding material comes up close to the window sash or door units, covering most of the old frame material.

Supplemental Exterior Work

Your re-siding work, when completed, will probably leave certain jobs undone. Even if prefinished siding material was used, there will probably be some portion of the home's exterior that wasn't covered and will need fresh paint or enamel. If using unfinished wood or plywood sidings, they will need finishing with a suitable paint or stain.

Some supplemental jobs that can be done conveniently at this time have been mentioned. Recovering eave soffits and fascia is one possibility. Essentially the same materials used for re-siding can also be applied on soffits, rakes, and fascia. Roof drainage downspouts (sometimes called leaders) will have been removed to allow the re-siding application, and gutters will have been removed to permit fascia recovering. It is thus a convenient time to also replace the complete roof drainage system. Improving the efficiency of window and door areas, by the use of storm sash, replacement windows and installation of thermal-type entrance doors, has also been discussed in earlier chapters. These make up the principal supplementary jobs in connection with re-siding work. With the possible exception of replacement windows, each of the foregoing jobs is well within the average capabilities of the homeowner and do-it-yourselfer. One more complicated type of home improvement under consideration here is the replacement of home entrance facilities by tearing out the old and installing new concrete stoops or wood entrance decks.

The intent in this chapter will be to provide the homeowner with guidance information, rather than an in-depth discussion of details. All of these supplemental jobs affect the home's appearance. Since improving the home's exterior appearance is usually one of the key motivating factors in re-siding, it makes sense for the owner to update one or more of these other exterior areas.

Exterior Finish Work

Caulking First, there's the matter of caulking. If your re-siding material has been metal or vinyl, the necessary caulking may have been done prior to placement of window/door and building corner trim. If the re-siding material was wood or wood-based, the probability is that an all-over caulking job needs to be done wherever the new siding meets or butts to window/door and building corner trim.

There's a baffling array of caulking materials available, nearly all in cartridge form for easy application by means of a caulking gun. Here are a few tips garnered from a recent study of caulking-sealing materials by Consumers Union:

- durability is improved if the gap or joint is first primed and the caulking later painted over;
- silicone-rubber sealers are not suitable for house exterior work because they cannot be painted;
- butyl and latex brands tend to be thin, hard to tool, and likely to sag;
- acrylic-latex caulkings were among the top-rated brands with good characteristics after aging.

A little practice in the rear of the home will result in the ability to apply the caulking material in a steady even flow. In cutting off the sealed nozzle tip, use side-cutting pliers and cut at a slight angle at about ½ inch in from the tip, so that the nozzle opening is close to ¼ inch.

Continuous beads of caulking should be applied around all siding joints with window and door frames and with building corner trim members. In using a caulking gun, hold it to the surface and pull it, don't push. Hold at about a 45° angle to the surface and use care when stopping to pause, or the caulking material will continue to flow out of the nozzle. Seal the nozzle with tape between periods of usage. After applying the bead, it can be "tooled" even smoother and more uniformly by using a piece of cloth over your index finger, the cloth being regularly dipped into a solvent that is suitable for the type of caulking material being used.

Latex paint and enamel For homeowner application, the readily available acrylic latex paints and enamels are probably the best buy. These paints are often called "water-thinned" paints, but this is misleading. Unlike the old oil-base materials which often had to be thinned before use, these newer paints are ready to use as they come from the can and virtually never need thinning. And the water base advantage is in the clean-up. Brushes and rollers can be cleaned under a faucet or in a water pail.

The acrylic latex paints come in flat or semigloss types and in gloss enamels. Paint suppliers are usually able to custom-mix these paints to whatever exact color you choose from a very wide assortment of color samples.

"All raw wood should be primed before painting." You've heard that old saw before. Then, when you visit paint stores, look around for a choice of different primers. You will see primers for lacquers; you will see enamel undercoaters; and, if you look hard, you may find a "house paint primer." The reason for this sometimes apparent shortage of priming paints is most painters do not use a special priming material; the paint itself is used as a primer. In fact, proper terminology would be to refer to the initial coat of paint as "the prime coat." Then, the second coat is called the "topcoat."

If you're repainting old exterior material, be sure you clean its surfaces first with a detergent washing; a film of grime accumulates on every house. Don't paint over old, loose paint flakes; wire brush these areas down first. On previously painted surfaces that are relatively light-toned in color, choose your new paint for hiding power as well as durability. It's worth paying some extra money for a paint that will completely cover and hide the previous paint in one coat.

Besides the easy water clean-up possible with the acrylic latex paints, look also for these further advantages in selecting a brand:
• flow-on application not showing brush marks;
• fast-drying, but also easy brush-over that allows you to stop and then restart painting without revealing the stopping point or edge;
• washability without powdering or flaking.

Stains Usually used on rustic, rough-sawn or heavily textured unfinished plywoods, hardboards and fiberboards, exterior house stains are considerably different from the very thin wood stains used in cabinet and furniture work. House stains come in standard and heavy-bodied types. They contain pigments, not to enhance woodgrains but to provide a general color tone to the wall. The standard types are sometimes referred to as semitransparent stains but the heavy-bodied stains are pigmented to a greater degree and will be fairly opaque. Both types penetrate below the wood surface and thus provide more durability and require less frequent restaining.

Wood shingles and shakes are excellent re-siding materials for stain finishes. Although these cedar materials can be left to weather, they do so non-uniformly and may be subject to undesirable natu-ral staining by water runoff. Application of prepared house stains not only minimizes such damage but also gives a more exact and uniform color tone to the surface.

The newer house stains being marketed are of the acrylic latex type, and offer the same easy clean-up with water that the paints and enamels do.

Eaves and Roof Drainage

Many re-roofing contractors also do roof gutter work. This is true partly because deteriorated roof edges must be repaired before new roofing materials can be applied. And the repair work will often necessitate gutter removal. Frequently, the gutters are in such bad shape, they can't be removed without damaging them further or they're not worth trying to put back up again.

If your reguttering is done by a roofing contractor, it's likely to be done in the traditional manner. The new gutter material will be galvanized steel or aluminum, left unfinished on its surface; it will need to be painted. The eave or fascia distances will have been measured and the gutter lengths soldered or brazed together to form a single house-length unit lifted into position by three or four men. This is the more troublesome but perhaps less costly way to go. The other route is with gutter and downspout materials and accessories that come with a factory-finish and employ special joining accessories to eliminate soldering or brazing. This is the route that many re-siding specialists are taking. And the reason is logical—the prefinished gutter-downspout materials are being made and distributed by the manufacturers of aluminum and vinyl sidings. And most of these producers also supply materials for recovering eave soffits and roof fascia boards, the other areas along roof edges which are among the first to need maintenance work.

Soffit-fascia renewal If this recovering work is to be done, it will precede reguttering work. All re-siding specialist applicators do soffit-fascia work. They will probably suggest that it be done, since they are preparing an estimate for re-siding your home. Most such applicators will also do fascia-

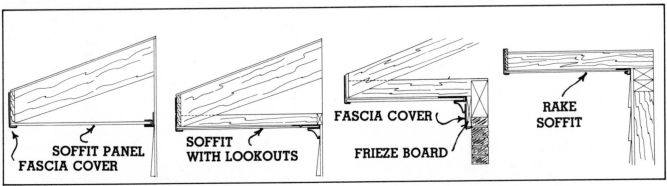

SOFFIT PANEL
FASCIA COVER

SOFFIT WITH LOOKOUTS

FASCIA COVER
FRIEZE BOARD

RAKE SOFFIT

soffit applications without re-siding, a situation that occurs frequently with homes having solid masonry or masonry veneer exterior walls.

On existing houses, the soffit-fascia work using aluminum or vinyl prefinished materials is essentially a recovering job. The soffit panels are identical or very similar to vertical siding panels. Formed and prefinished moldings are usually applied to conceal the panel edges at both the house and eave sides. However, house conditions vary in these details. Frieze boards may or may not have to be removed. Along the house wall, the new trim member may be a J-channel or a piece of custom-fitted trim. In some cases, where space permits, a frieze-runner is used. At the eave or fascia side, J-channel trim is normal, but sized to permit covering by the bottom of the fascia material.

Installation of aluminum or vinyl materials on soffits and fascia boards is quite simple, particularly on 1-story homes or houses having eaves at the same level and few dormers. It is, on such homes, well within the capabilities of do-it-yourselfers.

Where sheet materials such as plywood, hardboard, or fiberboard are being used for re-siding work, homeowners may prefer to make use of these materials as soffit replacements or recoverings. Some sheet material manufacturers offer special soffit materials, sometimes complete with ventilator openings. Vent-type soffit panels are also available in the prefinished aluminum and vinyl types.

Soffit-fascia work, regardless of the type material used, may be considerably easier if done as a replacement rather than a recovering job. Old soffit material can be removed quite easily. It very often cannot be firmly renailed because of moisture-damaged spots. The same may apply to fascia boards. Rotting can easily occur on the back of these boards as well as on rafter ends. The homeowner is much better off removing these old materials, repairing the lookout framing or rafter ends, applying a new fascia board, and then proceeding with the soffit and fascia materials.

Gutters and downspouts The prefinished aluminum, steel, and vinyl products are superior to the traditional installation of unfinished galvanized or aluminum gutters and downspouts. The reason is the finish. Galvanized comes to the job site with an oil film on its various surfaces. The same is true of aluminum, although the film is not so heavy. In either case, the seller or the applicator of the unfinished galvanized or aluminum will tell you to wait before painting. They'll explain that these products need a bit of weather exposure to oxidize their surfaces and make them suitable for painting. Don't believe them. In order to have your paint hold to the still-oily-after-weathering surface, you'll have to wash down the surfaces with a solvent. Then, a cleaner to remove the solvent residue. And for paint durability, this should probably be done on the inside of the gutter as well as the visible face. Otherwise rusting of galvanized or pitting of aluminum is encouraged.

Don't infer from the above that the prefinished products won't eventually be subject to similar damage. But it will take longer and you won't have the regular job of repainting and repair that seems to occur on most reguttering jobs that use raw metals plus paint. The emphasis above is on gutters; leaders or downspouts are much less prone to deterioration, though they may be just as troublesome in holding paint finishes.

There are differences, one manufacturer to another, in the prefinished aluminum and vinyl roof drainage products. But the similarities are much greater than the differences. The brief description below will apply to the various brands of either type of product. Remember, the key features of these roof drainage systems is their elimination of soldering or brazing, allowing installation with just simple hand tools.

Aluminum gutters are usually offered in a standard 0.027-inch gauge and a heavy-duty 0.032-gauge material. Downspouts are standard-sized 2¼ x 3¼-inch rectangular tubes formed from 0.02-inch gauge material. The gutters in both aluminum and vinyl are 5-inch boxtype, having what is referred to as an "OG" face or front edge shape.

A choice of gutter hangers is offered in order to fit various roof-edge conditions. Included in the choice will usually be a strap hanger that nails to the roof sheathing under the last shingle course, a bracket-hanger that nails to the fascia board, and 7 inch-long spikes with slender tube-like ferrules. The spikes are driven through the top outer edge of the gutter, through a ferrule and into the fascia and rafter width. A few companies have another hanging alternative—a roof edge strip with a hook-type edge on its bottom; the top edge of the gutter is also hooked and engages the bottom of the roof-edge strip for continuous support. Supplementary hanging brackets are still needed with this system.

Downspout outlets in these prefinished gutters are usually factory-prepared gutter sections with the outlet fitting premounted. Gutter lengths and outlet sections connect with each other through the use of slip-joint connectors. A sealing material is used with the slip-joints. In the case of vinyl guttering, the joining may use sealing cement plus pop rivets. These fastening devices are convenient tools also for working with aluminum gutters to obtain stronger joints than is possible with sealer alone.

For turning roof corners, both inside and outside corner accessory gutter units are available. Gutter material lengths vary somewhat but the common unit lengths are 10, 16, and 21 or 22 feet. Downspouts come in 10-foot lengths and two types of downspout elbows are used in making downspout connections. Gutter endcaps tap on where guttering ends. Other accessory items vary, but generally include strainers for downspout outlets, downspout fastening brackets and finish touch-up paint.

Replacement windows This subject was dealt with briefly as a key method for saving money on fuel bills. Allow a slight digression: many homeowners still successfully solve window problems by adding storm sash. The old single- or double-pane storm window glazed in a wood frame and hung on the outside of the prime window often offers a good tight fit and performs satisfactorily. The trouble is that with many older homes, the prime windows no longer offer much protection against air leakage, which diminishes the effect of adding storm windows.

Where the motivation is to eliminate changing storm windows and screens twice annually, the newer types of double- and triple-track storm-screen units are feasible. However, these flexible, self-storing storm windows, at $30-35 each for anodized-finished units and $45-50 each for baked enamel finish, can run into real money.

To spend as much as $50 per window to minimize heat losses/gains and save the work in changing storms/screens is fine. But it leaves the home with the old sash still on the inside where you have to put up with their poor operation, their poor appearance,

and the almost-annual touchup of glazing putty or paint. So, it frequently makes better economic sense to spend 10 to 25 percent more for good quality replacement windows, and have done with it.

This has become true only in recent years, as a few companies have begun to specialize in producing window units that can be custom-fitted to allow relatively easy insertion into old window frames. The windows are assembled in the plant according to exact measurements made at the home and included with the window order. This custom-fitting process is one in which a home-improvement contractor is usually involved; he does the measuring and ordering, then handles the installation work. There is a possibility that if you as a homeowner can demonstrate your ability to pay for the replacement windows, the manufacturer may ship them direct to you for do-it-yourself installation.

Replacement window installation on typical old double-hung windows of small to moderate size is a reasonably simple procedure. Moldings, stops, and old sash are removed, then the sash weights or balances. Parting stops are pried out and old moldings saved for reuse. Sash are removed from the replacement unit and the new window frame is trial-fitted to see the need for shims. Then, caulking

Below are indicated the various parts and connections that comprise the Kaiser system. This is the company's heavy-duty "Mark XXXII" line using .032 gauge aluminum. All parts have a durable plastic prefinish which will minimize the need for maintenance for many years.

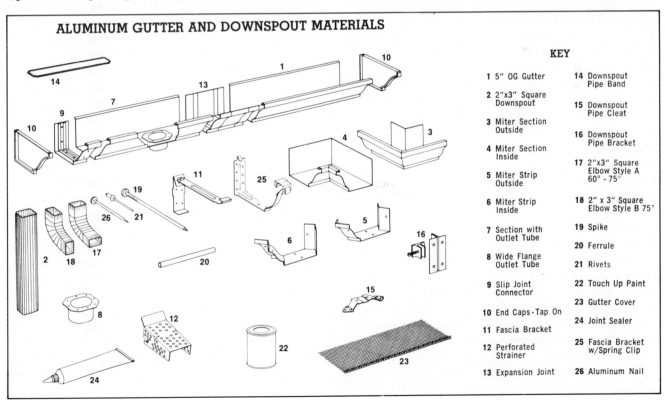

ALUMINUM GUTTER AND DOWNSPOUT MATERIALS

KEY

1 5" OG Gutter	14 Downspout Pipe Band
2 2"x3" Square Downspout	15 Downspout Pipe Cleat
3 Miter Section Outside	16 Downspout Pipe Bracket
4 Miter Section Inside	17 2"x3" Square Elbow Style A 60° - 75°
5 Miter Strip Outside	18 2" x 3" Square Elbow Style B 75°
6 Miter Strip Inside	19 Spike
7 Section with Outlet Tube	20 Ferrule
8 Wide Flange Outlet Tube	21 Rivets
9 Slip Joint Connector	22 Touch Up Paint
10 End Caps - Tap On	23 Gutter Cover
11 Fascia Bracket	24 Joint Sealer
12 Perforated Strainer	25 Fascia Bracket w/Spring Clip
13 Expansion Joint	26 Aluminum Nail

Simplified installation without the need for soldering explains the advantages of the new gutter-downspout kit-type systems now available in either aluminum or rigid vinyl. In the photo above, finishing connections *are being put on a gutter-downspout installation that used Kaiser Aluminum parts. Aluminum siding and soffit materials seen in the photo were also supplied by Kaiser.*

is applied against the blind stop at the window head and jambs. The window frame is reinserted, pushed against the blind stops and screw-fastened near the top. Alignment screws are then adjusted and the lower screws are fastened. Sash are installed, frame caulked, moldings replaced.

Obviously, the above paragraph is oversimplified. But don't fail to take into account the time factor as well as the replacement window cost factor. Timewise, a replacement window will take you less than an hour to install after you've done one, or thoroughly familarized yourself with the procedure. And this amount of time could very well be less than the time you need to repair or maintain the old window.

In replacement windows, you can obtain all the usual benefits provided by up-to-date storm sash: insulating properties, smooth operation, factory finishes that go years with only minimum upkeep. But there are additional advantages with some better quality replacement windows. These include a thermal-break window construction that minimizes conduction transfer or heat gains/losses. And for win-

dow cleaning ease, replacement units are available with removable sash or with tilt-in sash permitting cleaning from inside the home.

A somewhat comparable situation holds for replacement entrance doors. As indicated previously, entrance modernization is often part of exterior renewal work, and there are complete prehung entrance door units available to provide a fairly easy installation. Complete prehung units or just single replacement doors can be bought with a thermal-type construction using an insulative core material and pre-engineered tight-fitting edges. These replacement doors eliminate the need for storm doors, although the latter can still add to thermal efficiency. More common today than just storm doors are the so-called "combination" doors in which glazed storm panels are interchangeable with screen panels. Factory prefinishes minimize care to keep such doors in good shape. Typical cost of good-quality combination doors, not including installation, is in the $80-125 range. Figure about double that for replacement door-and-frame units.

Index